my **revision** notes

OCR AS
BIOLOGY

Frank Sochacki

Hodder Education, an Hachette UK company, 338 Euston Road, London NW1 3BH

Orders
Bookpoint Ltd, 130 Milton Park, Abingdon, Oxfordshire OX14 4SB
tel: 01235 827827
fax: 01235 400401
e-mail: education@bookpoint.co.uk
Lines are open 9.00 a.m.–5.00 p.m., Monday to Saturday, with a 24-hour message answering service. You can also order through the Hodder Education website: www.hoddereducation.co.uk

ISBN 978-1-4441-7967-5

First printed 2012
Impression number 5 4 3 2 1
Year 2017 2016 2015 2014 2013 2012

Typeset by Datapage (India) Pvt. Ltd.
Printed in India

Hachette UK's policy is to use papers that are natural, renewable and recyclable products and made from wood grown in sustainable forests. The logging and manufacturing processes are expected to conform to the environmental regulations of the country of origin.

P2182

Get the most from this book

Everyone has to decide his or her own revision strategy, but it is essential to review your work, learn it and test your understanding. These Revision Notes will help you to do that in a planned way, topic by topic. Use this book as the cornerstone of your revision and don't hesitate to write in it — personalise your notes and check your progress by ticking off each section as you revise.

☑ Tick to track your progress

Use the revision planner on pages 4 and 5 to plan your revision, topic by topic. Tick each box when you have:

- revised and understood a topic
- tested yourself
- practised the exam questions and gone online to check your answers and complete the quick quizzes

You can also keep track of your revision by ticking off each topic heading in the book. You may find it helpful to add your own notes as you work through each topic.

My revision planner

Unit F211 Cell, exchange and transport

	Revised	Tested	Exam ready
1 Cell structure			
7 Microscopes	☐	☐	☐
8 Cells	☐	☐	☐
2 Cell membranes			
14 Structure of cell membranes	☐	☐	☐

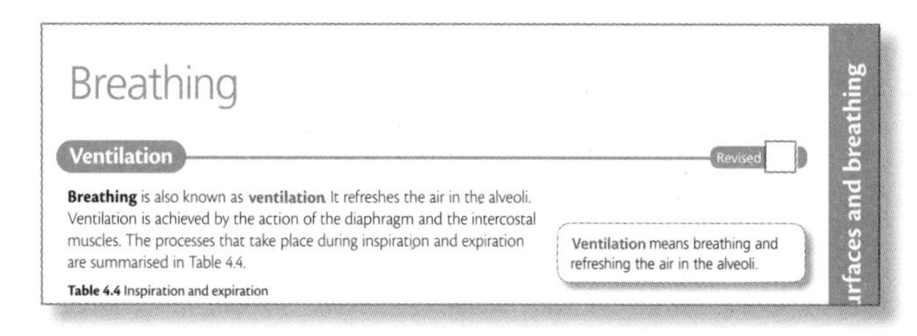
Breathing

Ventilation — Revised

Breathing is also known as **ventilation**. It refreshes the air in the alveoli. Ventilation is achieved by the action of the diaphragm and the intercostal muscles. The processes that take place during inspiration and expiration are summarised in Table 4.4.

Table 4.4 Inspiration and expiration

Ventilation means breathing and refreshing the air in the alveoli.

Features to help you succeed

Examiner's tips and summaries

Expert tips are given throughout the book to help you polish your exam technique in order to maximise your chances in the exam.

The summaries provide a quick-check bullet list for each topic.

Typical mistakes

The author identifies the typical mistakes candidates make and explains how you can avoid them.

Now test yourself

These short, knowledge-based questions provide the first step in testing your learning. Answers are at the back of the book.

Definitions and key words

Clear, concise definitions of essential key terms are provided on the page where they appear. Key words from the specification are highlighted in bold for you throughout the book.

Revision activities

These activities will help you to understand each topic in an interactive way.

Exam practice

Practice exam questions are provided for each topic. Use them to consolidate your revision and practise your exam skills.

Online

Go online to check your answers to the exam questions and try out the extra quick quizzes at **www.therevisionbutton.co.uk/myrevisionnotes**

My revision planner

Unit F211 Cells, exchange and transport

Unit F212 Molecules, biodiversity, food and health

Exam practice answers and quick quizzes at **www.therevisionbutton.co.uk/myrevisionnotes**

Countdown to my exams

6–8 weeks to go

- Start by looking at the specification — make sure you know exactly what material you need to revise and the style of the examination. Use the revision planner on pages 4 and 5 to familiarise yourself with the topics.
- Organise your notes, making sure you have covered everything on the specification. The revision planner will help you to group your notes into topics.
- Work out a realistic revision plan that will allow you time for relaxation. Set aside days and times for all the subjects that you need to study, and stick to your timetable.
- Set yourself sensible targets. Break your revision down into focused sessions of around 40 minutes, divided by breaks. These Revision Notes organise the basic facts into short, memorable sections to make revising easier.

Revised

4–6 weeks to go

- Read through the relevant sections of this book and refer to the examiner's tips, examiner's summaries, typical mistakes and key terms. Tick off the topics as you feel confident about them. Highlight those topics you find difficult and look at them again in detail.
- Test your understanding of each topic by working through the 'Now test yourself' questions in the book. Look up the answers at the back of the book.
- Make a note of any problem areas as you revise, and ask your teacher to go over these in class.
- Look at past papers. They are one of the best ways to revise and practise your exam skills. Write or prepare planned answers to the exam practice questions provided in this book. Check your answers online and try out the extra quick quizzes at **www.therevisionbutton.co.uk/myrevisionnotes**
- Use the revision activities to try different revision methods. For example, you can make notes using mind maps, spider diagrams or flash cards.
- Track your progress using the revision planner and give yourself a reward when you have achieved your target.

Revised

One week to go

- Try to fit in at least one more timed practice of an entire past paper and seek feedback from your teacher, comparing your work closely with the mark scheme.
- Check the revision planner to make sure you haven't missed out any topics. Brush up on any areas of difficulty by talking them over with a friend or getting help from your teacher.
- Attend any revision classes put on by your teacher. Remember, he or she is an expert at preparing people for examinations.

Revised

The day before the examination

- Flick through these Revision Notes for useful reminders, for example the examiner's tips, examiner's summaries, typical mistakes and key terms.
- Check the time and place of your examination.
- Make sure you have everything you need — extra pens and pencils, tissues, a watch, bottled water, sweets.
- Allow some time to relax and have an early night to ensure you are fresh and alert for the examination.

Revised

My exams

AS Biology Unit F211

Date: ....

Time: ....

Location:......................................

AS Biology Unit F212

Date: ....

Time: ....

Location:......................................

1 Cell structure

Microscopes

Types of microscope

Revised

Three types of microscope are commonly used, as shown in Table 1.1.

Table 1.1 Types of microscope

Type of microscope	Magnification	Resolution	Use
Light	1000–2000×	50–200 nm	Viewing **tissues** and **cells**
Scanning electron	50 000–500 000×	0.4–20 nm	Viewing the surface of cells and **organelles** Providing depth/three-dimensional images
Transmission electron	300 000–1 000 000×	0.05–1.0 nm	Detailing organelles (**ultrastructure**)

Examiner's tip

For the resolution of light and transmission electron microscopes, just remember the figure 0.2 and provide the unit appropriate to each type of microscope, e.g. 0.2 μm for light microscopes and 0.2 nm for transmission electron microscopes.

Now test yourself

Tested

1 Explain why a light microscope will not usually magnify images to greater than 1500×.
2 How can you tell the difference between an image created by a scanning electron microscope and one created by a transmission electron microscope?

Answers on p.108

Advantages and disadvantages

Light (optical) microscopes allow us to see living things, but unfortunately their **resolution** is limited.

Resolution is the ability to distinguish two separate points that are close together.

One advantage of using a scanning or transmission electron microscope is that the resolution is better, which means it is worth magnifying the image more as the image will show more detail. The main disadvantage is that the sample must be dried and is therefore dead. This may affect the shape of the features seen. Another disadvantage is that the image is in black and white only, but colours may be added later using computer graphics. These images are called false colour electronmicrographs. All electron microscopes are large and expensive.

Revision activity

Draw a table to show the advantages and disadvantages of the three types of microscope. The table should have three columns headed *Type of microscope*, *Advantages* and *Disadvantages*.

Magnification and resolution

Revised

Magnification is the ratio of the image size to the actual object size (size of image/size of object). Resolution is the ability to distinguish between two objects that are close together — the ability to provide detail in the image.

Magnification is the ratio of the image size to the object size.

Use the magnification triangle

where:

I is the **I**mage size

A is the **A**ctual size of the object

M is the **M**agnification

This means that:

$$I = A \times M$$

or:

$$A = \frac{I}{M}$$

or:

$$M = \frac{I}{A}$$

Revision activity

Make up an acronym to help you remember the magnification triangle.

Examiner's tip

Almost every biology examination paper is likely to include a calculation. This will often be to do with magnification or image size. You need to be able to calculate the magnification, but you also need to be able to manipulate the formulae.

If you can remember how to do the calculations, these are easy marks.

Now test yourself Tested

3 Using the magnification triangle $M = \frac{I}{A}$, rearrange the letters to give the formulae for I and for A.

4 Explain the relationship between micrometers (μm) and nanometers (nm).

Answers on p. 108

Staining

Revised

Staining is the application of coloured stains to the tissue. It makes objects visible in light microscopes and increases contrast so that the object can be seen more clearly.

Stains are often specific to certain tissues or organelles. In an electron microscope, the stains are heavy metals or similar atoms that reflect or absorb the electrons. In a transmission electron microscope, the heavy metal covers the organelles whereas in a scanning electron microscope the heavy metals cover the whole surface.

Revision activity

Draw a mind map to show the reasons for staining cells and tissues. Include the names of any stains you may have used such as methylene blue and iodine.

Cells

Organelles

Revised

Cells are the basic unit of living organisms. All **eukaryotic cells** share a similar basic structure containing membrane-bound organelles. Each organelle, whether membrane-bound or not, has its own function within the cell, as shown in Table 1.2.

Now test yourself Tested

5 Explain why organelles such as mitochondria do not always look the same size and shape.

Answer on p. 108

Revision activity

From memory, make a list of all the membrane-bound organelles and note one function next to each. Make a separate list of the organelles that are not membrane-bound.

Table 1.2 Organelles and their functions

Organelle	Function	Diagram
Centrioles	Involved in the organisation of the microtubules that make up the cytoskeleton Form the spindle used to move chromosomes in nuclear division	
Chloroplasts	Site of photosynthesis	
Cilia	Small hair-like extension of cell surface membrane containing microtubules Large numbers work in synchronised fashion Able to move whole organism or to move fluid (mucus) across a surface	
Cytoskeleton	A network of microtubules and microfilaments Provides support and structure for the cell Enables movement of organelles inside the cell Enables movement of the whole cell	
Flagella	Large extension of cell surface membrane containing microtubules (in eukaryotes) Able to beat to enable locomotion or move fluids	
Golgi apparatus	Modifies proteins made in the ribosomes Often adds a carbohydrate group Repackages proteins into vesicles for secretion	
Lysosomes	Small vacuoles containing hydrolytic or lytic enzymes	
Mitochondria	Site of aerobic respiration	
Nuclear membrane/ envelope	The nuclear envelope separates the genetic material from the cytoplasm	
Nucleus	Controls the cell activities Contains the genetic material (chromosomes) The nuclear pores allow molecules of mRNA to pass from the nucleus to the ribosomes in the cytoplasm	
Nucleolus	Assembles the ribosomes	
Ribosomes	Site of protein synthesis	
Rough endoplasmic reticulum (RER)	Holds the ribosomes Provides a large surface area for protein synthesis	
Smooth endoplasmic reticulum (SER)	Associated with the synthesis, storage and transport of lipids and carbohydrates	

Secretion of proteins

Revised

The organelles in a cell work together to achieve the overall function of that cell. Many of the organelles are involved in the production and secretion of **proteins**. The sequence of events always follows the same course:

1. mRNA leaves the nucleus via the nuclear pores.
2. It is used by the ribosomes on the rough endoplasmic reticulum to construct a protein.
3. The protein travels in a vesicle to the Golgi apparatus.
4. The vesicle is moved by the cytoskeleton, possibly using tiny protein motors that 'walk' along the microtubules using them as a track.
5. The Golgi apparatus modifies the protein (often adding a carbohydrate group) and repackages it into a vesicle.
6. This vesicle is moved to the cell surface or plasma membrane.
7. The vesicle fuses with the membrane to release the protein from the cell.

Revision activity

Draw a flow diagram of the sequence of events leading to the secretion of a protein.

Examiner's tip

Always remember to say that the plasma membrane is involved in secretion and that the vesicle fuses to this membrane.

Prokaryotic and eukaryotic cells

Revised

There are two types of cell: prokaryotic and eukaryotic cells. The features of each type are given in Table 1.3.

Table 1.3 Summarising the differences and similarities between prokaryotic and eukaryotic cells

Feature	Prokaryote	Eukaryote
Size	Smaller — typically less than 10 μm long and 1–2 μm wide	Larger — typically larger than 10 μm in diameter
Nucleus	No	Yes
Membrane-bound organelles	No	Yes
Ribosomes	Yes, 18 nm in size	Yes, 22 nm in size
Chromosomes	A single loop of DNA, no histones	DNA associated with proteins (histones)
Flagellum	Some cells have a flagellum. It has a very different structure	Some cells have a flagellum with 9 + 2 structure of microtubules

Animal and plant cells

Revised

The main differences between animal and plant cells are summarised in Table 1.4.

Table 1.4 Comparing animal and plant cells

Feature	Animal	Plant
Size	Variable, but typically 20 μm	Variable, but typically 40 μm
Cell wall	No	Yes — cellulose
Organelles	Ribosomes, ER, mitochondria, centrioles, nucleus, lysosomes	As animals, but contain chloroplasts and starch grains. No centrioles

Revision activity

- Make a list of the organelles found in an animal cell.
- Make a list of the organelles found in a plant cell.
- List the features of a bacterial cell that are also seen in plant cells.

Exam practice

1 A student prepared a slide of onion epidermis using the stain methylene blue.

(a) Explain the advantages of staining a tissue such as onion epidermis. [2]

(b) The student drew the following cell. Calculate the magnification of the image. [2]

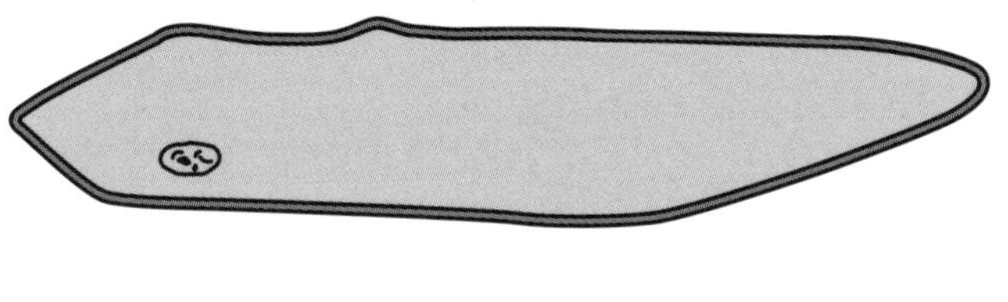

10 µm

(c) What feature in the student's diagram indicates that the cell is from a eukaryote? [1]

2 **(a)** Complete the following table to show organelles and their functions. [4]

Organelle	Function
Mitochondrion	
Chloroplast	
	Modify proteins
	Manufacture proteins

(b) What is the role of the nuclear pores? [2]

(c) Describe how the structure of a mitochondion is adapted to its function. [2]

(d) Suggest why lytic enzymes are held inside lysosymes. [2]

(e) Describe how organelles are moved around the cell by the cytoskeleton. [2]

Answers and quick quiz 1 online

Online

Examiner's summary

By the end of this chapter you should be able to:

- ✔ State the resolution and magnification that can be achieved by a light microscope, a scanning electron microscope and a transmission electron microscope.
- ✔ Explain the difference between magnification and resolution.
- ✔ Explain the need for staining samples for use in light and electron microscopy.
- ✔ Describe and interpret drawings and photographs of eukaryotic cells and recognise the nucleus, nucleolus, nuclear envelope, rough and smooth endoplasmic reticulum, Golgi apparatus, ribosomes, mitochondria, lysosomes, chloroplasts, plasma (cell surface) membrane, centrioles, flagella and cilia.
- ✔ Outline the function each organelle has in the cell.
- ✔ Recall that most organelles consist of membranes, and that the nucleus, mitochondria and chloroplasts have two membranes.
- ✔ Explain that membranes inside cells are distinct from the cell surface membrane that surrounds the cytoplasm.
- ✔ Outline the sequence of events leading to the secretion of a protein from the cell, including the interrelated roles of organelles.
- ✔ Explain that the cytoskeleton provides mechanical strength to cells, aids transport within cells and enables cell movement.
- ✔ Describe the differences between plant and animal cells.
- ✔ Describe the differences between prokaryotic and eukaryotic cells.

2 Cell membranes

The structure of cell membranes

Membranes in the cell

Revised

The **plasma (cell surface) membrane** surrounds the plasma. It keeps the contents of the cell separate from its surroundings, and limits what molecules can enter and leave the cell. It enables the cell to communicate with other cells through the process of **cell signalling**.

Other membranes in the cell make up organelles, which compartmentalise the cell. These membranes allow the cell to separate cell processes, which enables each process to occur in a specialised area of the cell. For example, all the enzymes involved in one process can be kept close together and other processes do not interfere. Concentration gradients can be formed across the membrane.

Examiner's tip

Always remember to refer to the outer membrane of the cell as the plasma membrane or cell surface membrane.

Revision activity

- Draw a diagram of a cell to show all the membranes inside the cell surface membrane.
- From memory, write a list of four functions of cell membranes.

Now test yourself

Tested

1 Explain how a concentration gradient can be built up.

2 Suggest why a compartmentalised cell is more efficient than one that is not compartmentalised.

Answers on p. 108

The fluid mosaic model

Revised

The **fluid mosaic model** (Figure 2.1) is used to describe the molecular arrangement of all membranes in a cell. A fluid mosaic membrane consists of:

- a bilayer of **phospholipid** molecules
- **cholesterol** which regulates the fluidity of the membrane, making it more stable
- **glycolipids** and **glycoproteins** which function in cell signalling or cell attachment
- **protein** molecules that float in the phospholipid bilayer. Some proteins are partially embedded in the bilayer — these are called extrinsic proteins. Other proteins span the bilayer — these are called intrinsic proteins. Some proteins float freely in the bilayer whereas others may be bound to other components in the membrane or to structures inside the cell

The **fluid mosaic model** is the basic structure of membranes in a cell.

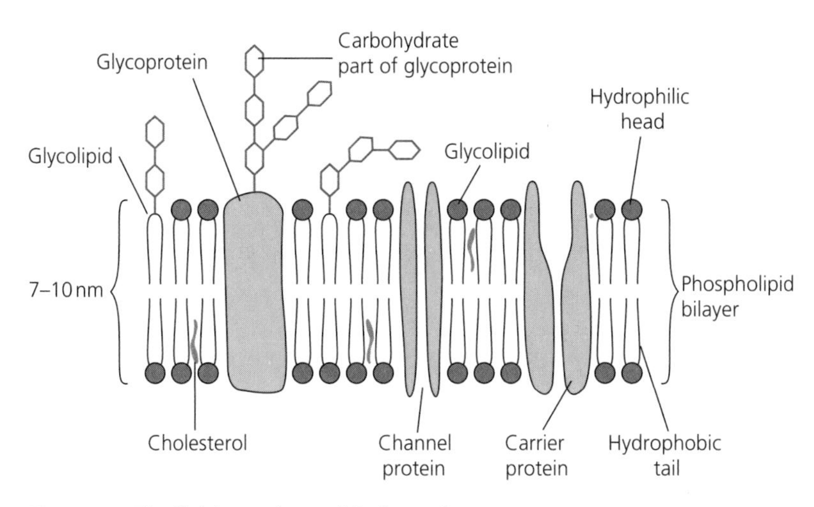

Figure 2.1 The fluid mosaic model of membrane structure

Components

Each component of the membrane has a specific role to play in the membrane.

Phospholipids

Phosopholipids provide a barrier that limits the movement of certain substances into and out of the cell, or into and out of the organelles, so the membrane is partially permeable. Small, fat-soluble substances can dissolve into the phospholipid bilayer and diffuse across the membrane. Water-soluble molecules and ions cannot easily dissolve and diffuse across the membrane. Some small molecules, such as water, may diffuse across slowly, but most require special transport mechanisms.

Cholesterol

Cholesterol fits between the tails of the phospholipid molecules. It inhibits the movement of the phospholipids, reducing the fluidity of the membrane. It also has the effect of holding the phospholipid tails together, providing some mechanical stability. Cholesterol makes the membrane less permeable to water and other substances such as ions.

Glycoproteins and glycolipids

The carbohydrate group attached to the protein or lipid molecule always has a specific shape, which can be used to recognise the cell — to identify it as 'self' or 'foreign'. Antigens on cell surfaces are usually glycoproteins or glycolipids.

The specific shape also allows the molecule to be used as a specific receptor site, as the shape may be complementary to the shape of other molecules in the body. Such complementary shapes can be used as binding sites for attachment of cell-signalling molecules such as **hormones**. If the correct **membrane-bound receptor site** is not present, the cell cannot respond to a hormone in the blood. These binding sites are also sites where drugs can bind to the cell, as many medicinal drugs either enhance or inhibit the action of the hormones that normally bind here. Another use for such binding sites is for cell attachment — the cells of a tissue bind together to hold the tissue together.

Revision activity

From memory, draw a small diagram of a part of a cell surface membrane to show the variety of molecules involved in its structure.

Examiner's tip

Never describe the cholesterol as increasing the rigidity of the membrane.

Proteins

Proteins have a variety of functions, such as enzymatic activity and cell signalling. However, many functions involve moving substances across the membrane. For example, some proteins may form:

- pores that allow the movement of molecules that cannot dissolve in the phospholipid bilayer
- carrier molecules that allow facilitated diffusion
- active pumps

Now test yourself Tested

3 Explain how the membrane can be selectively permeable.

Answer on p. 108

Revision activity

- Draw a diagram of a membrane including all the molecules involved in its structure and annotate it to describe what each component does.
- Draw a table listing each *Membrane component* in the left-hand column and its *Function* in the right-hand column.

The effect of temperature

Revised

Membranes are partially permeable, fluid and stable at normal body temperature. If temperature increases, the molecules gain kinetic energy and move about more. This increases the permeability of the membranes to certain molecules. Any molecules that diffuse through the phospholipid bilayer will diffuse more quickly. This is because as the phospholipids move about, they leave temporary gaps between them, providing space for small molecules to enter the membrane.

If temperature increases further, the phospholipid bilayer may lose its mechanical stability (it may melt) and the membrane becomes even more permeable. Eventually, the proteins in the membrane will denature. This will further damage the structure of the membrane and it will become completely permeable.

Typical mistake

Many candidates describe the membrane as denaturing, but it is only the proteins that denature.

Cell signalling

Revised

Cell signalling is the way in which cells communicate within an organism to coordinate the activities of that organism. It is achieved by releasing chemicals from one cell, which move around the body to a target cell. The target cell surface membrane has a receptor that is specific to the signal molecule. The shape of the receptor is complementary to the shape of the signal molecule. Nerve synapses and hormones are both examples of cell signalling.

Drugs can be made that fit the receptors on certain cells. For example, asthmatics take Salbutamol, which fits the receptors on smooth muscle in the airways to cause relaxation.

Now test yourself Tested

4 Explain the importance of complementary shapes in cell signalling.

Answer on p. 108

Revision activity

Describe how a cell surface membrane may be different from the membranes inside a cell.

Examiner's tip

Always say that the shape of the signal molecule is complementary to the shape of the receptor molecule.

Transport of substances across membranes

Passive transport

Revised

Passive transport is the movement of molecules that does not need metabolic energy in the form of ATP. It uses energy in the form of the kinetic (movement) energy. It only occurs when there is a concentration gradient and when molecules move down a concentration gradient.

Note that because molecules move randomly, some molecules may move in the 'wrong' direction — so you should describe passive transport as the net movement of molecules down their concentration gradient.

Passive transport can occur in three forms:

- **Diffusion** (Figure 2.2) —the net movement of molecules away from a concentrated source. This may occur across a membrane if the molecules are fat-soluble or if they are small and can fit between the phospholipids in a membrane.
- **Facilitated diffusion** (Figure 2.2) — diffusion across a membrane that is helped by a protein in the membrane. The protein could be a pore protein (which may be gated) or it could be a carrier protein.
- **Osmosis** — the net movement of water molecules across a partially permeable membrane. Water molecules move down their **water potential gradient** (i.e. from an area of higher **water potential** to an area of lower water potential).

Passive transport is the movement of molecules without the use of metabolic energy in the form of ATP.

Diffusion is the net movement of molecules down a concentration gradient.

Facilitated diffusion is diffusion that is aided by a protein in the membrane.

Osmosis is the movement of water from a region of higher water potential to a region of lower water potential.

Water potential gradient is the difference between the water potential in one place compared to another.

Water potential is a measure of the tendency of water molecules to move from one place to another.

Now test yourself

Tested

5 Explain why charged ions must be transported by facilitated diffusion rather than by simple diffusion.

Answer on p. 108

Examiner's tip

Don't describe or explain osmosis in terms of 'water concentration' as this can be confused with the concentration of solutes. Always use the term *water potential.*

The rate of diffusion

Diffusion occurs without using metabolic energy. It relies on the kinetic energy of the molecules. The rate of diffusion is affected by a number of factors:

- Temperature — a higher temperature gives molecules more kinetic energy. At higher temperatures the molecules move faster, so the rate of diffusion increases.
- Concentration gradient — more molecules on one side of a membrane (or less on the other) increases the concentration gradient. This increases the rate of diffusion.
- Size of molecule — small molecules or ions can move more quickly than larger ones. Therefore, they diffuse more quickly than larger ones.

- Thickness of membrane — a thick barrier creates a longer pathway for diffusion. Therefore, diffusion is slowed down by a thick barrier or membrane.
- Surface area — diffusion across membranes occurs more rapidly if there is a greater surface area.

Active transport

Revised

Active transport (Figure 2.2) moves molecules against their concentration gradient and uses membrane-bound proteins that change shape to move the molecules across the membrane.

Active transport is the movement of molecules using metabolic energy in the form of ATP.

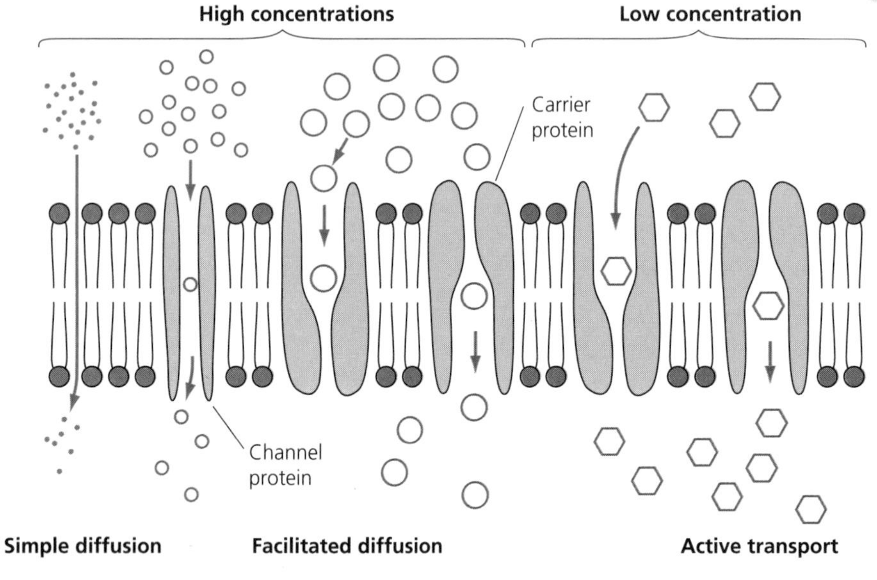

Figure 2.2 Diffusion and active transport

Transport proteins

Revised

Transport proteins are proteins that help transport substances across membranes. They include carrier proteins, which may move molecules across in both passive and active transport by changing shape. Channel proteins carry out passive transport and provide a channel through which molecules can move.

Bulk transport

Revised

Bulk transport is the movement of molecules through a membrane by the action of vesicles. **Endocytosis** is the formation of vesicles by the plasma membrane, which moves molecules into the cell. **Exocytosis** is fusion of vesicles with the plasma membrane, which moves molecules out of the cell. Bulk transport uses metabolic energy.

Revision activity

Draw a mind map to show how a range of different substances pass through cell membranes.

Now test yourself

Tested

6 Explain why proteins pass through membranes by bulk transport.

Answer on p. 108

Water potential

Revised

Water potential is a measure of the ability of water molecules to move freely.

Pure water has a water potential of zero. As solutes (sugars or salts) are added to a solution, the water potential gets lower. Therefore, a salt solution has a water potential of below zero, i.e. a negative potential.

Water molecules will move from a solution with a higher water potential to a solution with a lower (more negative) water potential. Therefore, water molecules always move down their water potential gradient.

Typical mistake

Some candidates confuse water potential and water potential gradient. A gradient can only exist between two places.

Osmosis

Revised

A cell placed in pure water will have a lower (more negative) water potential than the surrounding water. There will be a water potential gradient from high outside the cell to lower inside the cell. As a result, water molecules will enter the cell.

A cell placed in a strong salt solution will have a higher (less negative) water potential than the surrounding solution. There will be a water potential gradient from higher inside the cell to lower outside the cell. As a result, water molecules will leave the cell.

Examiner's tip

It may be easier to describe osmosis in terms of water molecules moving from a less negative region to a more negative region.

Table 2.1 The effects of osmosis on animal and plant cells

	Solution	
Cell type	**Pure water**	**Strong salt**
Animal	An animal cell has no cell wall. The plasma membrane has no strength, so the animal cell will burst.	An animal cell has no cell wall. The cytoplasm will shrink and the cell will shrivel. Its appearance is known as crenated.
Plant	Water makes a plant cell turgid. The vacuole is full of watery sap and the cytoplasm pushes the plasma membrane out against the cell wall. The cell wall is strong and will stop the cell bursting.	A plant cell will lose its turgidity. It will become flaccid. If the water loss continues, the cell vacuole will shrink. The cytoplasm will also shrink and the plasma membrane pulls away from the cell wall. This is called plasmolysis.

Now test yourself

Tested

7 Explain why a plant cell will not burst when placed in pure water.

8 Using the terms *water potential* and *water potential gradient*, explain why a plant cell loses turgidity when placed in a strong salt solution.

9 In a plasmolysed plant cell, state what is found in the gap between the cell wall and the plasma membrane. Explain how it gets there.

Answer on p. 108

Exam practice

1 **(a)** Describe how small molecules such as water and carbon dioxide can cross a plasma membrane. [2]

(b) Describe how large molecules can be brought into a cell. [3]

2 **(a)** The diagram below shows four animal cells that touch each other. Each cell has a different water potential as shown by the figures. Draw arrows to show the direction in which water will move by osmosis from cell to cell. [4]

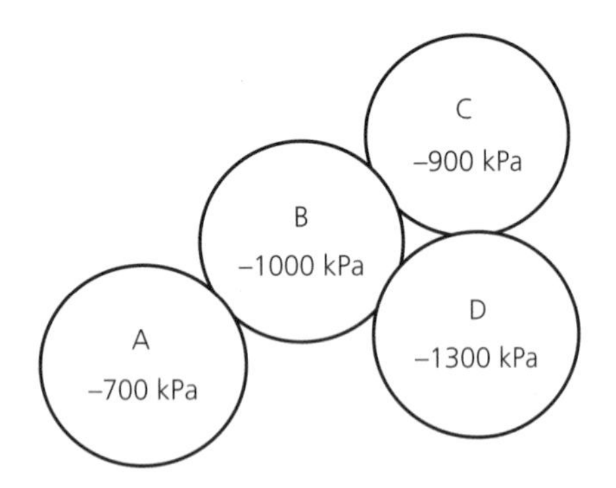

(b) Using the term *water potential*, explain the movement of water that would occur if cell C in the diagram were placed in pure water. [3]

3 Describe the functions of the following components of a cell membrane:

(a) cholesterol [1]

(b) a glycoprotein [2]

(c) phospholipids [2]

4 Use the idea of cell signalling to describe how the cell ensures that vesicles containing proteins can be transported to the correct organelle inside a cell. [4]

Answers and quick quiz 2 online

Online

Examiner's summary

By the end of this chapter you should be able to:

- ✔ Outline the roles of membranes in cells.
- ✔ Describe the fluid mosaic model of membrane structure.
- ✔ Describe the roles of the components in a membrane.
- ✔ Outline the effect of changing temperature on membrane permeability and structure.
- ✔ Explain the role of membrane-bound receptors in cell signalling.
- ✔ Understand how substances pass through membranes.

3 Cell division, cell diversity and cellular organisation

Cell division

Mitosis

Revised

Mitosis produces two genetically identical cells which are used for:

- **growth** of the organism
- **repair** of tissues
- replacement of old cells
- **asexual reproduction**

Mitosis is the division of a cell to produce two genetically identical daughter cells.

Asexual reproduction is the production of offspring from one parent and does not involve sex.

The cell cycle

Mitosis is a small part of the **cell cycle** (Figure 3.1). The remainder of the cycle (known as interphase) is used for copying the chromosomes and checking the genetic information. During interphase the cell also increases in size, produces new organelles and stores energy for another division.

Typical mistake

Some candidates lose marks because they suggest that mitosis is for the growth or repair of cells. This is not true — cells grow during interphase and mitosis repairs tissues not cells.

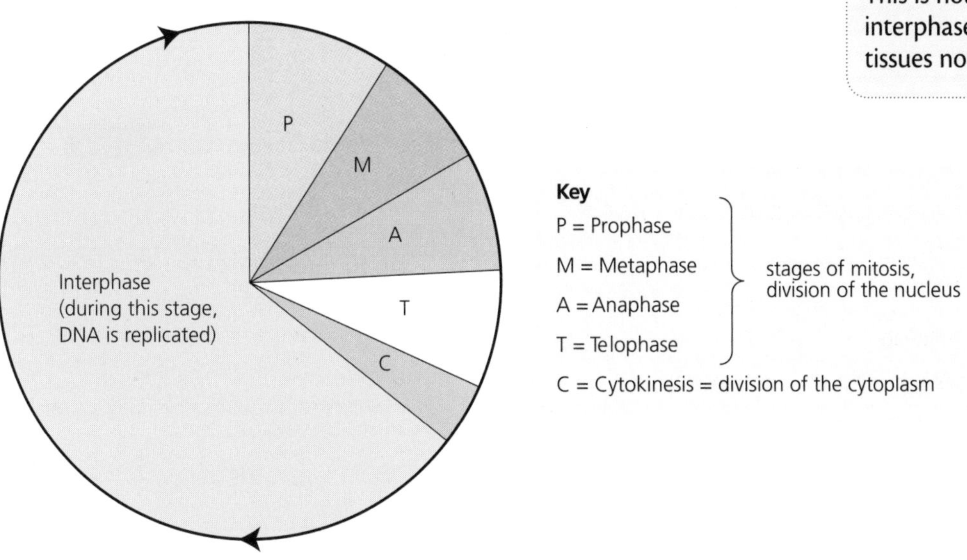

Figure 3.1 The cell cycle. Mitosis (PMAT) takes up about 5–10%, which is too short a time relative to interphase to show accurately on this diagram

Now test yourself

Tested

1 Explain why the cell must copy the DNA before mitosis.
2 Explain why the cell must produce new organelles and store energy during interphase.

Answers on p. 108

The stages of mitosis

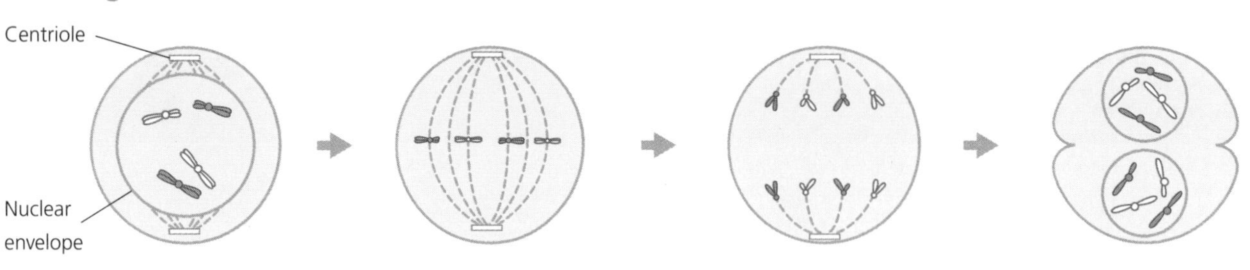

Prophase

- The start of mitosis
- Chromosomes shorten and thicken as DNA is tightly coiled
- Each chromosome is visible as two chromatids joined at the centromere
- Prophase ends as the nuclear envelope breaks into small pieces
- Centrioles organise fibrous proteins into the spindle

Metaphase

- Chromosomes are held on the spindle at the middle of the cell
- Each chromosome is attached to the spindle on either side of its centromere

Anaphase

- Chromatids break apart at the centromere and are moved to opposite ends of the cell by the spindle

Telophase

- Nuclear envelopes reform around the chromatids that have reached the two poles of the cell
- Each new nucleus has the same number of chromosomes as the original, parent cell
- The nuclei are genetically identical to each other

Figure 3.2 The stages of mitosis in an animal cell

Budding in yeast

Budding is a form of asexual reproduction that allows yeast cells to reproduce quickly when the conditions are suitable. The nucleus undergoes mitosis and one nucleus migrates to one side of the cell. The cell wall bulges and the nucleus moves into the bulge. The bulge increases in size and eventually constriction occurs to separate the new daughter cell from the parent.

Examiner's tip

Remember that the cell cycle and mitosis are continuous processes and the names of the stages reflect parts of a continuum.

Homologous chromosomes

Revised

An **homologous pair of chromosomes** (Figure 3.3) is a pair of chromosomes which have the:

- same shape and size
- centromere in the same position
- same genes in the same positions on the chromosomes

An **homologous pair of chromosomes** is a pair of chromosomes that carries matching genes.

Centromere

Three pairs of homologous chromosomes

Figure 3.3 Homologous chromosomes

Meiosis

Revised

Meiosis is an alternative form of cell division. Meiosis produces four cells that are:

- not genetically identical
- gametes
- haploid (contain half the normal number of chromosomes)

Typical mistake

Many candidates confuse meiosis with the process of fertilisation, which takes place after meiosis.

Examiner's tip

You do not need to know any details of meiosis.

Stem cells and differentiation

Stem cells

Revised

Stem cells are cells that are not specialised or differentiated. They maintain the capacity to undergo mitosis and differentiate into a range of cell types. **Differentiation** is the ability of a cell to specialise to form a particular type of cell. Stem cells are found in many animal tissues (e.g. **bone marrow**) and in the meristem tissues of plants. Bone marrow cells can differentiate into blood cells (**erythrocytes** and **neutrophils**). **Cambium** cells in plants can differentiate into **xylem** and **phloem**.

Now test yourself

3 Explain why cells that have already differentiated are unable to divide and produce a range of cell types.

Answer on p. 108

Tested

Cell specialisation

Revised

Cells are specialised in a number of ways, as shown in Table 3.1.

Table 3.1 Cell specialisation

Cell type and function	Specialisation	How that specialisation helps function
Erythrocytes (red blood cells) carry oxygen in the blood	Small and flexible	To fit through tiny capillaries
	Full of haemoglobin	To bind to the oxygen
	No nucleus	To allow more space for haemoglobin
	Biconcave shape	To provide a large surface area to take up oxygen
Neutrophils engulf and digest foreign matter or old cells	Flexible shape	To enable movement through tissues
	Lobed nucleus	To help movement through membranes
	Many ribosomes	To manufacture digestive enzymes
	Many lysosomes	To hold digestive enzymes
	Many mitochondria	To release the energy needed for activity
	Well developed cytoskeleton	To enable movement
	Membrane-bound receptors	To recognise materials that need to be destroyed
Sperm carry the paternal chromosomes to the egg	Tail (flagellum)	To enable rapid movement
	Acrosome	To help digest the egg surface
	Small	To make movement easier
	Many mitochondria	To release the energy needed for rapid movement
Epithelial cells act as surfaces	Often flat (**squamous**)	To cover a large area
	Often thin (squamous)	To provide a short diffusion distance
	May be **ciliated**	To move mucus
	May be cuboid	To provide a barrier
	Many glycolipids and glycoproteins in cell surface membrane	To hold cells together as a surface
Palisade cells for photosynthesis	Elongate	To fit many chloroplasts into the space
	Contain many chloroplasts	To absorb as much light as possible
	Show cytoplasmic streaming	To move the chloroplasts around
	Contain starch grains	To store the products of photosynthesis
Root hair cells absorb water and mineral ions from the soil	Long extension (hair)	To increase surface area
	Active pumps in cell surface membrane	To absorb mineral ions by active transport
	Thin cell wall	To reduce the barrier to movement of ions and water
Guard cells control the stomatal opening	Active pumps in cell surface membrane	To move mineral ions in and out of the cell to alter the water potential
	Unevenly thickened wall	To cause the cell to change shape as it becomes more turgid
	Large vacuole	To take up water and expand to open the stoma

Revision activity

Draw a detailed diagram of each cell type in Table 3.1 and annotate it to explain how the cell is specialised.

Tissues, organs and organ systems

Organisation

Revised

A **tissue** is a collection of cells that work together to perform a particular function. They may be similar to each other or they may perform slightly different roles. For example:

- ciliated epithelium contains ciliated cells that move mucus over their surface and goblet cells that produce the mucus
- xylem contains vessels that carry water. The cells are specialised in that they are joined end to end with no end walls. They have no cytoplasm or nucleus and contain bordered pits in their lignified walls
- phloem contains two types of cell — sieve tube elements, which have a thin layer of cytoplasm containing few organelles and no nucleus, and companion cells, which have dense cytoplasm and a large nucleus. The sieve tube elements and the companion cells are linked by many plasmodesmata

An **organ** is a collection of tissues working together to perform a common function.

An **organ system** is made up of two or more organs working together to perform a life function such as excretion or transport.

Revision activity

Write separate lists of all the tissues and organs in the human transport and gaseous exchange systems.

Now test yourself

4 Explain why forming tissues is more efficient than using individual cells to perform a task.

Answer on p. 108

Tested

Cooperation

Revised

A multicellular organism can become more efficient if the cells are specialised to perform different roles. However, the cells must be able to communicate with each other to ensure that each cell works with the others in a coordinated fashion. This is achieved through cell signalling. For example, the muscles can only work harder if the lungs and blood system supply more oxygen and remove more carbon dioxide.

Revision activity

Draw a mind map to show how cells cooperate with one another in a multicellular organism.

Exam practice

1 **(a)** Explain what is meant by the term *differentiation*. [2]

(b) Beta cells in the pancreas are specialised to produce the protein hormone insulin. Suggest how these cells may be specialised. [3]

(c) Diabetes is a disease in which the beta cells stop producing insulin. Suggest how stem cells could be used to cure this disease. [2]

(d) Explain the advantages of using stem cells from an embryo. [2]

2 **(a)** Define the terms *tissue* and *organ*. [4]

(b) Plant transport tissues are called xylem and phloem. Describe how cells are organised to form these tissues. In your answer you should use technical terms correctly. [5]

Answers and quick quiz 3 online

Online

Examiner's summary

By the end of this chapter you should be able to:

- ✔ Describe the cell cycle and the stages of mitosis.
- ✔ Explain the significance of mitosis for growth, repair and asexual reproduction.
- ✔ Understand the meanings of the terms *homologous chromosomes, stem cell, differentiation, budding, tissue, organ* and *organ system.*
- ✔ Describe and explain how certain cells are specialised for their function.
- ✔ Understand why cells in multicellular organisms need to specialise and cooperate.

4 Exchange surfaces and breathing

Exchange surfaces

Surface area to volume ratio

Revised

A living organism needs to absorb substances from its surroundings and remove waste products. This can only occur through its surface area. However, as an organism increases in size, its volume increases so it needs more from its environment. Unfortunately, its surface area does not increase as quickly as its volume, so the larger an organism gets, the more difficult it becomes to absorb enough substances over its surface. This can be demonstrated by considering a simple set of data:

- Assume that the organism is a cube with sides of length *l*.
- Its surface area is 6 × the area of one side or 6 × length squared ($6 \times l^2$).
- Its volume is length × length × length or length cubed (l^3).

Table 4.1 shows what happens to surface area, volume and **surface area to volume ratio** as an organism increases in size.

Table 4.1 The effects on an organism as it increases in size

Length of organism (*l*) (mm)	Surface area of organism ($6 \times l^2$) (mm^2)	Volume of organism (l^3) (mm^3)	Surface area to volume ratio
1	6	1	6
5	150	125	1.25
10	600	1000	0.6

We can see that as size increases:

- surface area increases
- volume increases, but more quickly than surface area
- surface area to volume ratio decreases

The surface area to volume ratio is the surface area of an organism divided by its volume. It is a key concept as the surface area must be able to provide sufficient oxygen through diffusion from the environment.

Revision activity

Sketch a graph of surface area to volume ratio plotted against body size for an amoeba, a mouse, a man and an elephant.

Significance

Single-celled organisms are small and have a large surface area to volume ratio. Their surface area is large enough for sufficient oxygen and nutrients to diffuse into the cell to provide all its needs, and for wastes to diffuse out.

As an organism increases in size, becoming **multicelluar**, its surface area to volume ratio gets smaller. **Diffusion** is too slow for the oxygen and nutrients to diffuse across the whole organism. The surface area is no longer large enough to supply all the needs of the larger body. Therefore, a **specialised exchange surface** is required, such as lungs in animals used for gaseous exchange.

Examiner's tip

Remember that:

- length, surface area and volume all have different units
- surface area to volume ratio has no units
- suitable units must always be included in any work involving figures

Now test yourself Tested

1. List the factors that affect the need for a specialised surface for gaseous exchange.
2. Explain why a single-celled organism such as an amoeba does not need a specialised surface for gaseous exchange whereas a large tree does.

Answers on p. 108

Typical mistake

Many candidates confuse surface area with surface area to volume ratio. An elephant has a large surface area but a small surface area to volume ratio.

Good gaseous exchange surfaces

Revised

A good **gas exchange surface** (Figure 4.1) must be able to exchange gases quickly enough to provide for the activity of the cells inside the organism. The surface has certain features, as shown in Table 4.2.

Squamous means flattened.

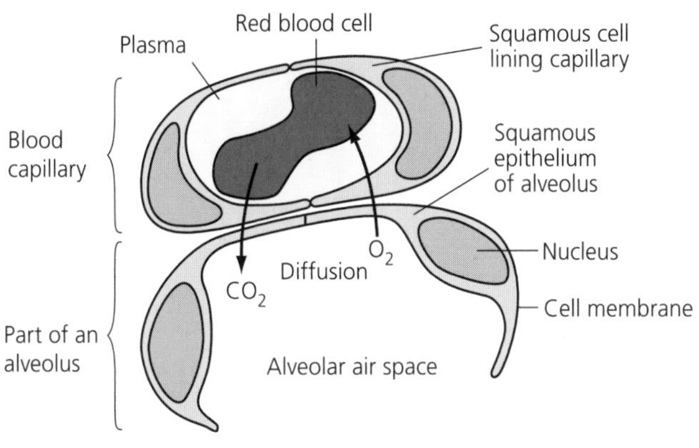

Figure 4.1 The gaseous exchange surface in the lungs

Typical mistake

Many candidates describe the lungs as having a 'thin cell wall' — they probably mean a 'wall of thin cells' or a 'thin wall of cells'. This sort of vague wording should be avoided. Describe the barrier as creating a short diffusion pathway because the cells are squamous.

Table 4.2 Features of good surfaces for gaseous exchange

Feature	Reason	In the lungs
Large surface area	To provide space for molecules of oxygen and carbon dioxide to pass	Lung epithelium folded to form numerous **alveoli** (singular: alveolus)
Thin barrier	To provide a short diffusion pathway	Lung epithelium and capillary endothelium are both made from **squamous** cells
Steep concentration gradient	To ensure molecules diffuse rapidly in the correct direction	Good supply of blood on one side and ventilation of the air sacs on the other side

The **concentration gradient** is the difference in concentration between two points. In the lungs the presence of a very thin barrier to diffusion helps to create a steep concentration gradient.

This steep concentration gradient is maintained by increasing the concentration of molecules on the supply side and reducing the concentration on the demand side. In the lungs this is achieved by good blood flow and ventilating the air spaces. Blood flow brings carbon dioxide to the lungs and removes **oxygen** whereas ventilation brings oxygen to the lung surface and removes **carbon dioxide**.

Revision activity

Draw a mind map to link the features of a good gaseous exchange surface to the rate of diffusion.

Now test yourself Tested

3. List the factors that affect the concentration gradient.

Answer on p. 108

Examiner's tip

Always remember to describe changes in the concentration of the gases in the blood or air sacs as it is the concentration gradient that drives diffusion.

The gaseous exchange system

Cells and tissues in the lungs

Revised

Figure 4.2 shows the structure of the human gaseous exchange system, consisting of the trachea and lungs (Figure 4.3).

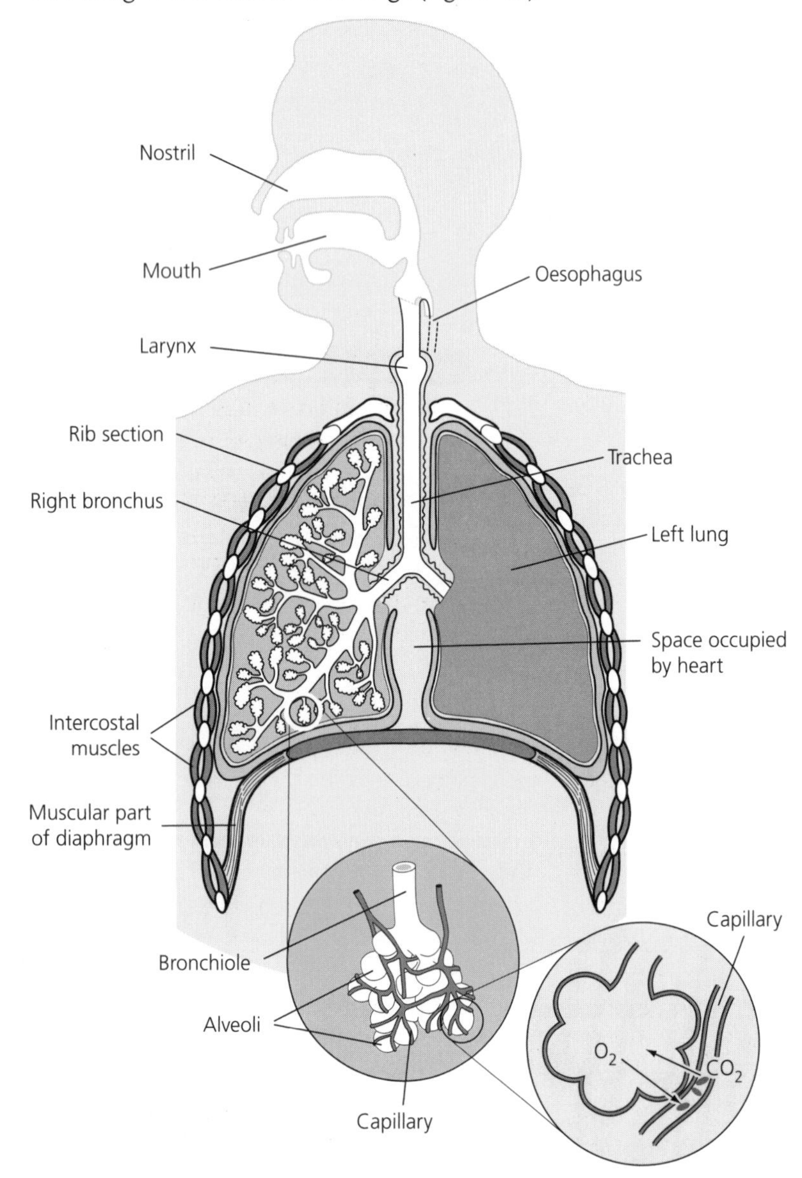

Figure 4.2 The gaseous exchange system, with details of the gaseous exchange surface formed by alveoli

Table 4.3 The distribution and function of structures in the lungs

Structure	Distribution	Function
Capillaries	Over surface of alveoli	To provide a large surface area for exchange
Cartilage	In walls of bronchi and trachea	To hold the airways open
Ciliated epithelium	On surface of airways	The cilia move or waft the mucus along
Elastic fibres	In walls of airways and over alveoli	To recoil to return the airway or alveolus to original shape. In alveolus this helps to expel air
Goblet cells	In ciliated epithelium	To produce and release mucus
Smooth muscle	In walls of airways	Contracts to constrict or narrow the airways
Squamous endothelium	Capillary wall	To provide a thin barrier to exchange — a short diffusion pathway
Squamous epithelium	Surface of alveoli	To provide a thin barrier to exchange — a short diffusion pathway

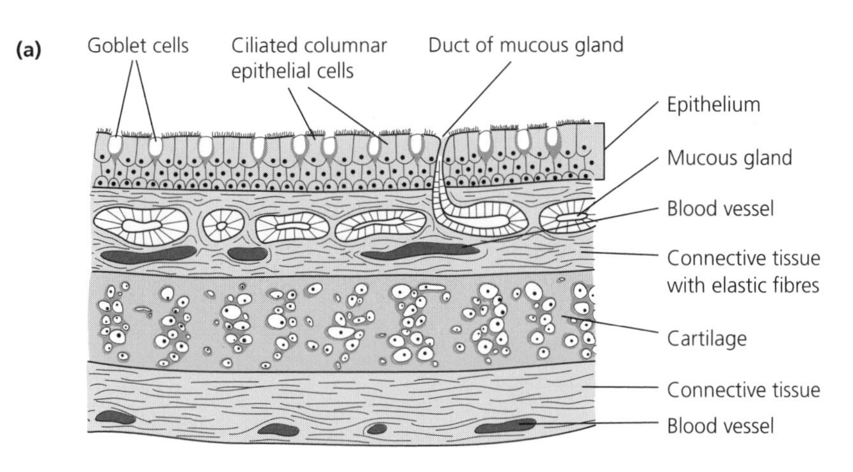

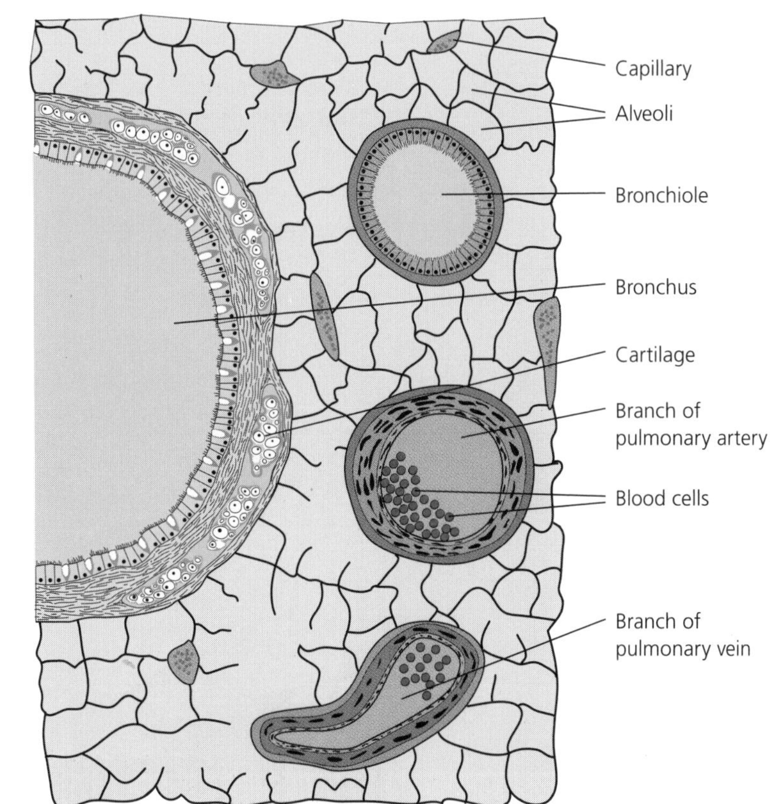

Figure 4.3 (a) Detail of the wall of the trachea (b) Distribution of tissues in the lungs

Typical mistake

Many candidates make the mistake of describing:

- cilia as 'hairs'
- cilia as trapping bacteria
- cilia as moving the mucus to the goblet cells, which then engulf the dust and pathogens in the mucus
- elastic tissue as contracting
- smooth muscle as providing a smooth surface to reduce friction

Revision activity

From memory, write a list of the tissues found in the lungs and describe what each does to help gaseous exchange.

Breathing

Ventilation

Revised

Breathing is also known as **ventilation**. It refreshes the air in the alveoli. Ventilation is achieved by the action of the diaphragm and the intercostal muscles. The processes that take place during inspiration and expiration are summarised in Table 4.4.

Ventilation means breathing and refreshing the air in the alveoli.

Table 4.4 Inspiration and expiration

Structure/feature	Inspiration (inhaling)	Expiration (exhaling)
Diaphragm	Contracts and moves downwards pushing organs down	Relaxes and is pushed up by organs underneath
Intercostal muscles	Contract to raise the rib cage up and outwards	Relax and allow the rib cage to fall
Volume change	Chest cavity increases in volume	Chest cavity reduces in volume
Pressure change	Pressure inside chest cavity reduces and falls below atmospheric pressure	Pressure inside chest cavity rises above atmospheric pressure
Air movement	Air is pushed into lungs by higher atmospheric pressure	Air is pushed out of lungs by higher pressure in alveoli

Revision activity

Draw a flow chart to describe inhaling and exhaling.

Examiner's tip

To achieve full marks, you will need to describe all the volume and pressure changes accurately.

Tidal volume and vital capacity

Revised

Tidal volume is the volume of air breathed in and then out in one breath. The tidal volume changes according to the needs of the body. At rest it is usually about 0.5 dm^3.

Vital capacity is the maximum volume of air that can be forced out after taking a deep breath. Vital capacity is typically 4.5 dm^3 in young men and 3.0 dm^3 in young women. Vital capacity can be increased through training. Singers and athletes often have a large vital capacity.

Tidal volume is the volume of air breathed in and then out in one breath.

Vital capacity is the maximum volume of air that can be forced out after a deep breath.

Using a spirometer

Revised

1. The subject should wear a nose clip to ensure that no oxygen escapes from the system and no additional air is added.
2. The subject breathes through the mouthpiece.
3. As the subject inhales, oxygen is drawn from the air chamber, which therefore descends.
4. As the subject exhales, the air chamber rises again.
5. Air returning to the air chamber passes through the canister of soda lime, which absorbs carbon dioxide.
6. The movements of the air chamber are recorded by a data logger or on a revolving drum.

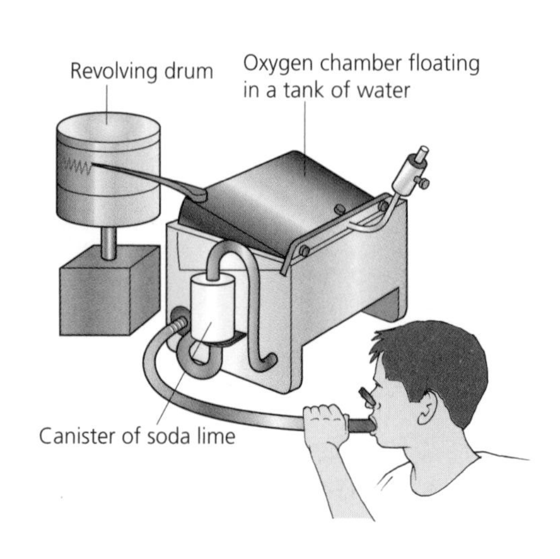

Figure 4.4 A spirometer

7 Tidal volume is measured simply by allowing the subject to breathe normally.

8 Vital capacity is measured by asking the subject to breathe out as deeply as possible.

Now test yourself Tested

4 Explain why the subject should wear a nose clip.

5 Explain the function of the soda lime and why it is essential.

Answers on p. 108

Using the spirometer trace

All measurements are taken from the spirometer trace (Figure 4.5). Always remember to measure at least three readings (if possible) and calculate a mean. **Breathing rate** is calculated by counting the number of peaks in one minute. **Oxygen uptake** is a little more difficult:

- As carbon dioxide is removed, the total volume in the air chamber decreases.
- The volume of carbon dioxide removed is shown by the difference in height of the last peak from the first peak during normal breathing.
- This can be assumed to equal the volume of oxygen used by the subject.
- Divide this volume by time taken in order to calculate the rate of oxygen uptake.

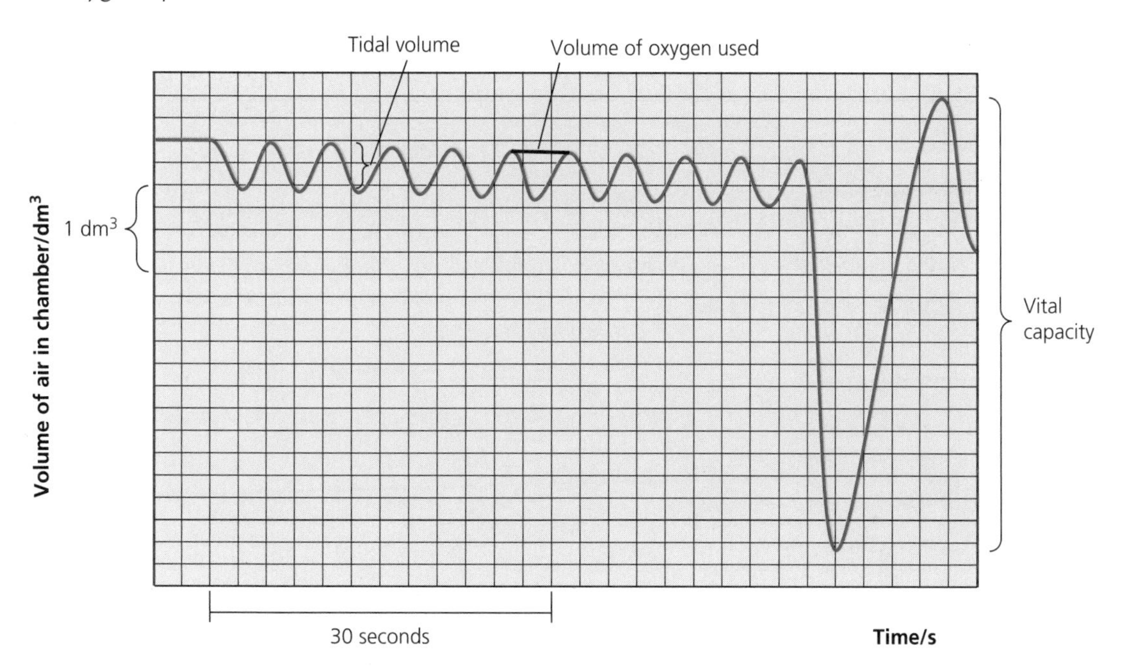

Figure 4.5 Measurements can be taken from the spirometer trace

Exam practice

1 **(a)** Explain why a large, active animal such as a mammal needs a specialised surface for gaseous exchange. [3]

(b) The following table shows how the surface area and volume of a sphere change as its size increases.

Diameter (cm)	Surface area (cm^2)	Volume (cm^3)	Surface area to volume ratio
2	50.3	33.5	1.5
5	314.2	523.7	0.6
10	1256.8	4189.3	

(i) Calculate the surface area to volume ratio of a sphere of 10 cm radius. Show your working. [2]

(ii) Describe the trend shown by the surface area to volume ratio as the size of the sphere increases. [2]

2 **(a)** State one function in the airways of each of the tissues listed below.

elastic tissue *ciliated epithelium* *smooth muscle* [3]

(b) Describe how ciliated cells and goblet cells work together to reduce the risk of infection in the lungs. [3]

(c) The alveoli walls contain elastic fibres. Suggest what may happen to the alveoli if the elastic fibres are damaged. [2]

(d) In asthmatics certain substances in the air cause the smooth muscle in the walls of the airways to contract. Suggest the effect this may have on the person. [2]

3 **(a)** Describe how you would use a spirometer to measure tidal volume. [3]

(b) Explain why the air chamber should be filled with medical-grade oxygen rather than air. [2]

(c) Describe two other precautions that should be taken when using a spirometer. [2]

Answers and quick quiz 4 online

Online

Examiner's summary

By the end of this chapter you should be able to:

- ✔ Understand the importance of surface area to volume ratios.
- ✔ Describe the features of a good gaseous exchange surface.
- ✔ Describe the features of the lungs that make them a good surface for gaseous exchange.
- ✔ Describe the distribution of tissues in the lungs and explain the role of each in an efficient organ of gaseous exchange.
- ✔ Outline the mechanism of breathing.
- ✔ Explain the terms *tidal volume* and *vital capacity*.
- ✔ Understand how a spirometer can be used to measure vital capacity, tidal volume, breathing rate and oxygen uptake.

5 Transport in animals

All living animal cells need a supply of oxygen and **nutrients** to survive. They also need to remove **waste** products such as carbon dioxide and urea so that they do not build up and become toxic. There are three main factors that affect the need for a transport system:

- size
- **surface area to volume ratio**
- level of activity

Multicellular animals are large organisms, so they need specialised exchange surfaces.

Typical mistake

Many candidates describe small organisms as having a large surface area rather than a large surface area to volume ratio, or they describe a large organism as having a small surface area.

Circulatory systems

Single and double circulatory systems

Revised

In a **single circulatory system**, blood flows through the heart once every time it goes around the body. **Fish** have a single circulatory system. The blood flows from the heart to the gills and then on to the body before returning to the heart:

heart → gills → body → heart

In a **double circulatory system**, blood flows through the heart twice for every circuit around the body. **Mammals** have developed a circulation that involves two separate circuits. One circuit carries blood to the lungs to take up oxygen. This is the pulmonary circulation. The other circuit carries the oxygen and nutrients around the body to the tissues. This is the systemic circulation. The heart is adapted to form two pumps, one for each circulation:

body → heart → lungs → heart → body

Typical mistake

Many candidates describe a single circulation as 'blood going through the heart once'. However, the blood must go around the body and return to the heart, so it is better to describe it as 'going through the heart once for every circuit of the body'.

Now test yourself

1 Explain why a double circulatory system is more efficient than a single circulatory system.

Answer on p. 108

Tested

Open and closed circulatory systems

Revised

Insects have an **open circulatory system**. In an open system:

- there is no separate tissue fluid
- blood circulates around the organs and cells
- pressure cannot be raised to help circulation
- circulation is affected by body movements
- oxygenated and deoxygenated blood mix freely

Fish have a **closed circulatory system**. In a closed system:

- blood is kept in vessels
- pressure can be maintained
- pressure can be higher
- flow can be faster
- flow can be directed to certain tissues or organs

Revision activity

Sketch an insect and a mammal. Inside each include a simple version of the circulatory system.

The heart

The **mammalian heart** is a muscular pump. It is divided into two sides. The right side pumps **deoxygenated blood** to the lungs to be oxygenated. The left side pumps **oxygenated blood** to the rest of the body. On both sides the action of the heart is to squeeze the blood, putting it under pressure and forcing it along the arteries.

Deoxygenated blood transports less oxygen and more carbon dioxide than oxygenated blood.

Oxygenated blood transports oxygen to the organs.

External features of the heart

Revised

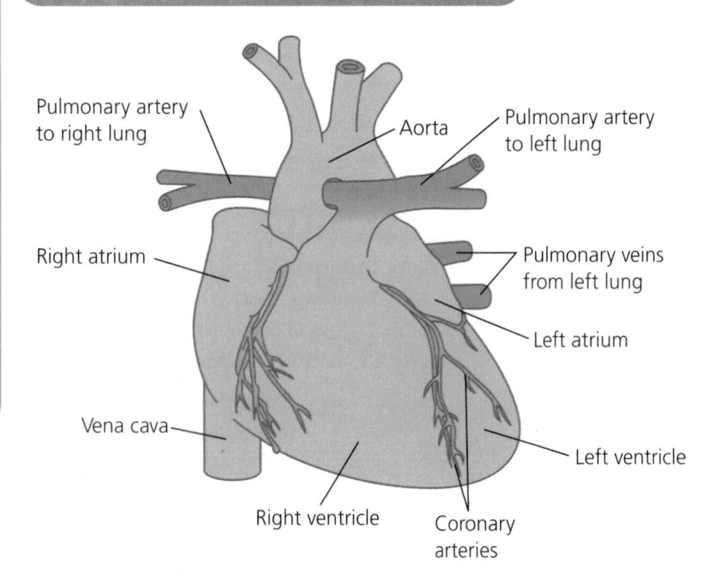

Figure 5.1 An external view of the heart

The muscle surrounding the two main pumping chambers (the ventricles) is dark red. Above the ventricles are two thin-walled chambers known as the atria (singular: atrium). These are much smaller than the ventricles. On the surface of the heart are coronary arteries. These carry oxygenated blood to the heart muscle itself. These arteries are important as the heart works hard continually. If they become constricted, blood flow to the heart muscle is restricted and this can reduce the delivery of oxygen and nutrients, such as fatty acids and glucose, causing a heart attack. At the top of the heart are the veins that carry blood into the heart and the arteries that carry blood out of the heart.

Revision activity

Draw a diagram to show the external features of the heart and annotate it with the names and functions of each feature.

Internal features of the heart

Revised

The heart is divided into four chambers: two atria and two ventricles.

The atria

The two upper chambers are **atria**. These receive blood from the major veins. Deoxygenated blood flows from the vena cava into the right

The atria are the small chambers at the top of the heart, which collect blood from the main veins.

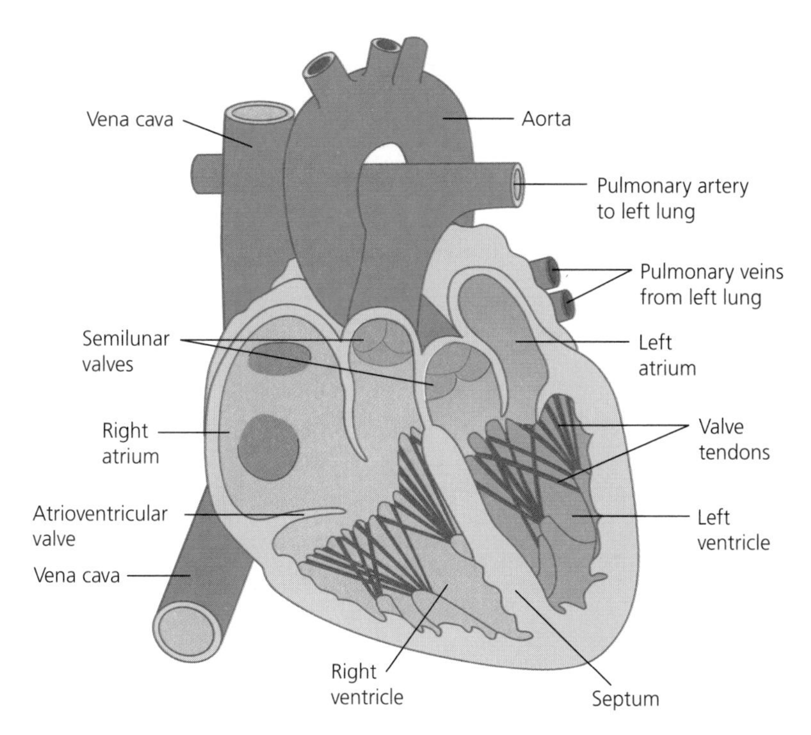

Figure 5.2 A vertical section through the heart

atrium. Oxygenated blood flows from the pulmonary vein into the left atrium. The atria have very thin walls as they do not need to create much pressure. Blood simply flows through the atria into the ventricles. When the ventricles are nearly full, the atrial walls contract just to completely fill the ventricles.

The ventricles are the larger lower chambers, which have thick walls to pump the blood out of the heart.

The **septum** is the muscle that separates the two ventricles.

The ventricles

The two lower chambers are the **ventricles**. Each has a thick muscular wall. The wall contracts to create pressure which pushes the blood into the arteries.

The right ventricle has walls that are thicker than the atrial walls. This enables it to pump blood out of the heart. The right ventricle pumps deoxygenated blood to the lungs. Therefore, the blood does not need to travel far. Also, the lungs contain many fine capillaries. The pressure of the blood must not be too high to prevent the capillaries in the lungs bursting.

Revision activity

Draw a diagram to show the internal features of the heart and annotate it with the names and functions of each feature.

The walls of the left ventricle are often two or three times thicker than those of the right ventricle. The blood from the left ventricle is pumped out through the aorta and needs sufficient pressure to propel it all the way to the extremities of the body. The pressure created by the left ventricle is typically measured at 110–120 mmHg.

Now test yourself

2 Explain why the left ventricle looks so much bigger than the right.

Answer on p. 109

Tested ☐

The ventricles are separated from each other by a wall of muscle called the **septum**. This ensures that the oxygenated blood in the left side of the heart and deoxygenated blood in the right side are kept separate.

Major arteries

Deoxygenated blood leaving the right ventricle flows into the pulmonary artery leading to the lungs. Oxygenated blood leaving the left ventricle flows into the aorta. This carries blood to a number of arteries that supply all parts of the body.

The cardiac cycle

Revised

It is important that the chambers of the heart all contract in a coordinated fashion. If the chambers were to contract out of sequence, this would lead to inefficient pumping. The sequence of events involved in one heart beat is called the **cardiac cycle** (Figure 5.3).

The **cardiac cycle** is the series of events in one heart beat.

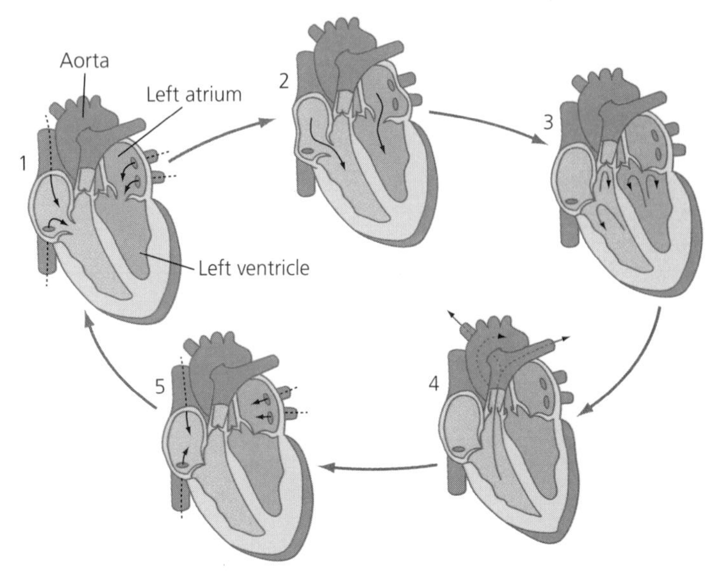

Figure 5.3 The cardiac cycle

1. Blood returns to the heart from the body (via the vena cava) and lungs (via the pulmonary vein). Both the atria fill at the same time. The **valves** between the atria and the ventricles (the **atrioventricular valves**) are open to allow blood to flow straight through into the ventricles.
2. Once the ventricles are nearly full, the **sinoatrial node (SAN)** initiates a new heartbeat. It creates a wave of excitation which spreads over the walls of the atria. The walls contract, pushing a little extra blood from the atria into the ventricles. The wave of excitation is stopped by a layer of non-conducting fibres between the atria and the ventricles. The wave of excitation can only pass through the **atrioventricular node (AVN)**, where it is delayed a little. This allows time for the ventricles to fill.
3. After the delay, the wave of excitation passes down the bundle of His in the septum between the ventricles. At the base of the septum, the bundle splits into separate fibres called **Purkine tissue** that carry the excitation up the walls of the ventricles, causing contraction from the base (apex) upwards. The walls of the two ventricles contract together. As the pressure rises, the atrioventricular valves are pushed shut which prevents blood re-entering the atria. The tendinous cords attached to the valves prevent them from inverting.
4. The blood pressure in the ventricles rises quickly until it exceeds the pressure in the aorta and pulmonary artery. This pushes the **semilunar valves** open and blood is pushed into the main arteries.

The **atrioventricular valves** lie between the atria and the ventricles.

Examiner's tip

Remember that it is the walls of the atria and ventricles that contract.

The **semilunar valves** lie at the entrance to the main arteries.

5 Once contraction is complete, the muscles relax and the elasticity of the walls causes recoil to return the ventricles to their original size and shape. This causes the pressure in the ventricles to drop quickly. When the pressure drops below the pressure in the main arteries, the semilunar valves are pushed shut, preventing re-entry of blood into the ventricles. When the pressure in the ventricles drops below the pressure in the atria, the atrioventricular valves are pushed open by the pressure in the atria. This allows blood to flow into the ventricles again.

Pressure changes during contraction

The changes in pressure in the heart can be represented by a graph (Figure 5.4). The important points are where one line crosses another — this is where the pressure in one chamber rises above that in another chamber, causing a valve to open or close.

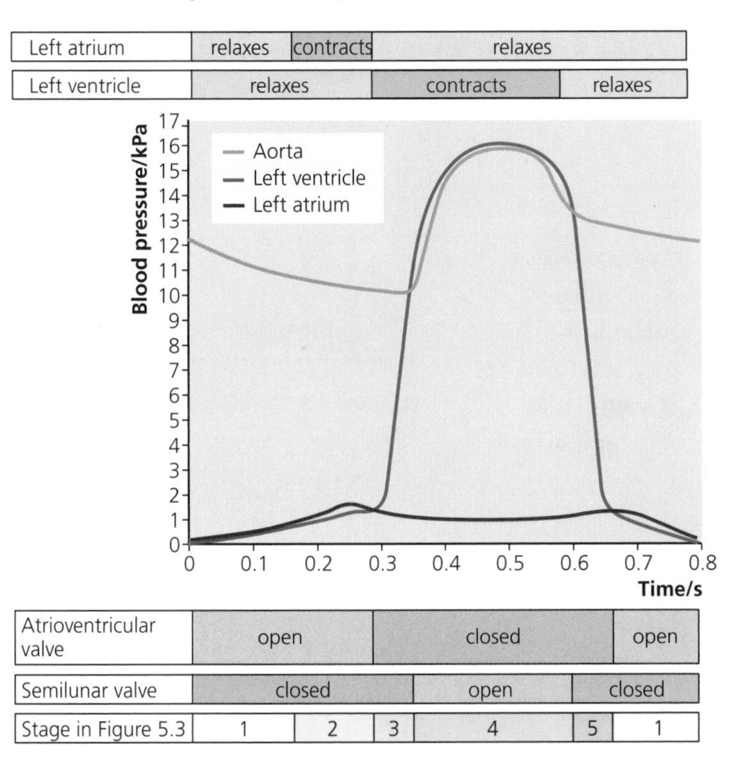

Figure 5.4 Contraction occurs in stages 2, 3 and 4 — this is the systole. Relaxation occurs in stages 5 and 1 — this is the diastole

Revision activity

Sketch an outline of the heart and draw arrows to show the direction of blood flow.

Now test yourself

3 Explain how the atrioventricular valves are opened and closed.

4 Explain why the electrical stimulation wave must be delayed at the AVN.

Answers on p. 109

Tested

Now test yourself

5 Describe and explain what happens at the point in Figure 5.4 where the line for the pressure in the left ventricle rises above the line for the pressure in the left atrium.

Answer on p. 109

Tested

Electrocardiograms

Revised

An **electrocardiogram (ECG)** (Figure 5.5) records the electrical activity of the heart. Wave P is the excitation of the atria. Wave QRS is the excitation of the ventricles. Wave T is associated with ensuring the muscles have time to rest.

Abnormal heart activity can often be identified by an abnormal ECG trace. The waves may be smaller, inverted or further apart.

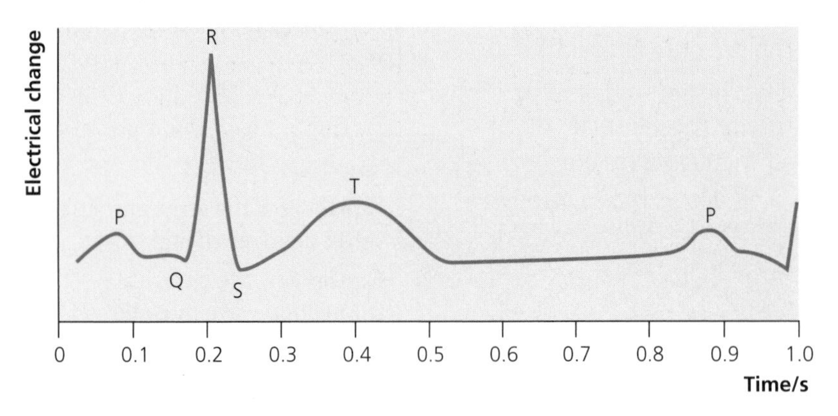

Figure 5.5 A normal ECG trace

Examiner's tip

Questions are likely to show you a normal trace and an abnormal trace and ask you to identify the differences.

Blood vessels

Arteries, veins and capillaries

Revised

Blood flows through a series of vessels. Each is adapted to its particular role in relation to its distance from the heart. All types of blood vessel have an inner layer or lining made of cells called the **endothelium**. This is a thin layer that is particularly smooth to reduce friction with the flowing blood. Figure 5.6 shows cross-sections of an **artery** and a **vein**. Table 5.1 compares the structure and functions of arteries, veins and **capillaries**.

The **endothelium** is the thin layer of cells that lines all blood vessels.

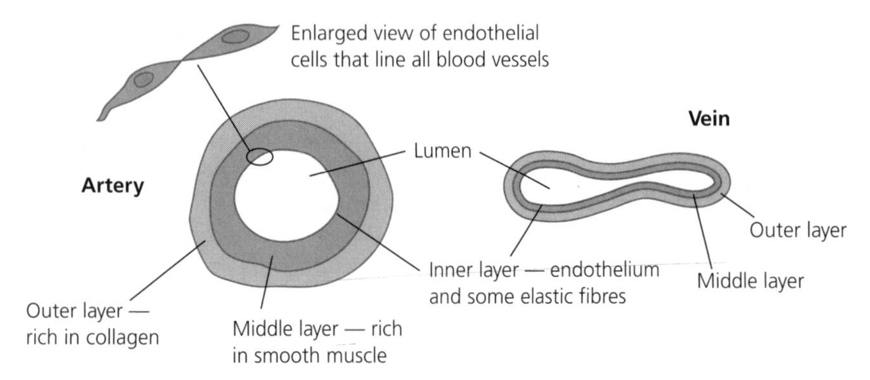

Figure 5.6 Cross-sections of an artery and a vein

Table 5.1 Comparing the structure and functions of arteries, veins and capillaries

Feature	Artery	Vein	Capillary
Function	Transports blood away from the heart	Transports blood back to the heart	Enables exchange of materials between the blood and tissue fluid
Thickness of wall	Thick	Thin	Very thin (one cell thick)
Components of wall	Endothelium lining surrounded by thick middle layer of elastic tissue and smooth muscle, then a thick outer layer rich in collagen	Endothelium lining surrounded by thinner layer of elastic tissue and smooth muscle, with a thin outer layer containing collagen	One layer of endothelium cells
Blood pressure	High	Low	Low
Presence of valves	No	Yes	No
Cause of flow	Pressure created by the heart, maintained by recoil of elastic tissues	Squeezing action of body muscles and valves to ensure correct direction	Pressure from action of the heart

Typical mistake

Many candidates describe the role of smooth muscle as 'creating a smooth surface to reduce friction', or they describe the action of the smooth muscle as 'pumping blood along the artery'. The smooth muscle is used to reduce the diameter of the vessel to maintain blood pressure or to reduce blood flow in a particular direction. Similarly, the elastic tissue is used to recoil the artery wall to maintain blood pressure.

Revision activity

Write a list of the features of the arteries that maintain blood pressure and another list of the features that enable the arteries to withstand high blood pressure.

Now test yourself Tested

6 Describe how the structure of an artery is adapted to its role of transporting blood at high pressure.

7 Describe how the structure of a vein is adapted to its role of transporting blood at low pressure.

Answers on p. 109

Blood, tissue fluid and lymph

Revised

Blood is the fluid found inside the blood vessels. It consists of:

- water-based plasma containing dissolved substances — oxygen, nutrients such as glucose and amino acids, lipoproteins, carbon dioxide (most transported as bicarbonate ions), other wastes such as urea, hormones, antibodies and plasma proteins
- red blood cells (erythrocytes), probably carrying oxygen
- white blood cells, such as neutrophils and lymphocytes
- platelets

Tissue fluid surrounds the body cells. It is the **plasma** that has been filtered out of the blood, so it contains all the dissolved elements of the blood except the cells, platelets and plasma proteins. These are too large to pass out of the blood vessels. There may be some phagocytic neutrophils in tissue fluid as these can change shape to squeeze out of the blood vessels.

Examiner's tip

Remember that tissue fluid is blood that does not contain blood cells or plasma proteins, but it does contain dissolved components.

Lymph is excess tissue fluid that is not returned to the blood vessel. Instead it is drained into the lymph vessels. These carry the fluid back to the circulatory system by a different route.

Lymph contains the same substances as tissue fluid but has less oxygen and glucose as these have been used by the cells. Lymphocytes produced in the lymph nodes may also be present.

Revision activity

Draw a table to compare the composition of blood, tissue fluid and lymph.

How is tissue fluid formed?

The walls of the capillaries are a single layer of endothelium cells. Fluid and dissolved substances in the fluid can squeeze between the endothelium cells (Figure 5.7).

At the arterial end of the capillary, the hydrostatic pressure created by the heart is still sufficient to squeeze fluid out of the capillary. The pressure of the blood is 4.4 kPa compared to just 1.1 kPa in the tissue fluid. Dissolved substances such as oxygen and nutrients pass out with the fluid.

Once the fluid has left the capillary, it is called tissue fluid. Oxygen and nutrients in the tissue fluid surrounding the body cells can diffuse into those cells. Waste products such as carbon dioxide can diffuse into the tissue fluid.

Some fluid is retained in the capillaries so that the cells continue to flow along. This is because the plasma proteins maintain a low (more negative) water potential. The water potential of the blood is –3.3 kPa. This water potential is lower than the water potential of the tissue fluid (–1.3 kPa). Therefore, water is drawn back into the capillary at the venous end of the capillary where the hydrostatic pressure has dropped significantly lower.

Carbon dioxide and other wastes are carried back into the blood as tissue fluid moves back into the capillary. Some of the tissue fluid is drained into the blind-ending lymph vessels to become lymph.

Examiner's tip

Don't worry too much about the figures — they have been included to help you understand the process. Just remember that at the arterial end of the capillary the hydrostatic pressure of the blood pushes fluid out of the capillary and at the venous end the difference in water potentials moves fluid back into the capillary.

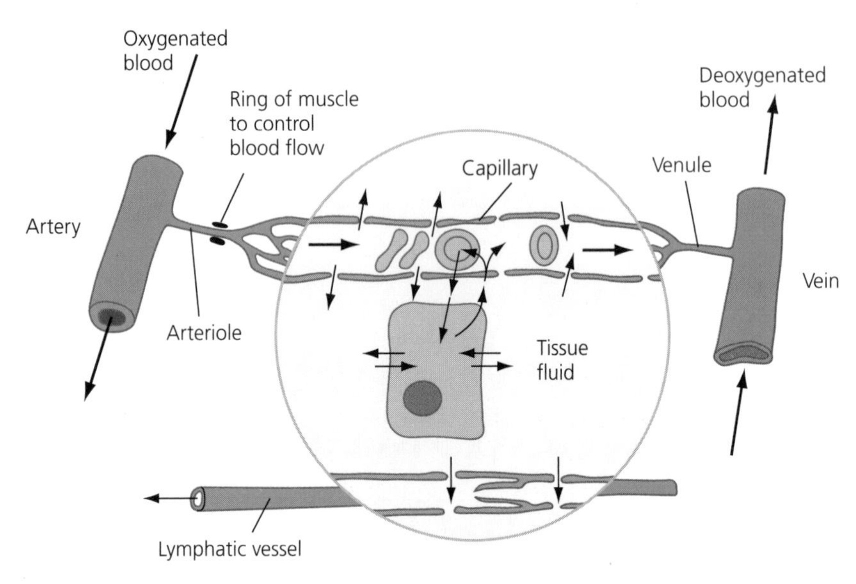

Figure 5.7 Blood flow through a capillary bed

Transport of oxygen and carbon dioxide

Transport of oxygen

Revised

Oxygen enters the blood in the lungs. Oxygen molecules diffuse into the blood plasma and red blood cells. Here, they associate with the **haemoglobin** (Hb) to form **oxyhaemoglobin**.

Haemoglobin is a complex protein with four subunits. Each subunit contains a haem group that contains a single iron ion (Fe^{2+}). This attracts and holds one oxygen molecule. The haem group is said to have an **affinity for** oxygen. Each haem group can hold one oxygen molecule, so each haemoglobin molecule can carry four oxygen molecules.

Haemoglobin is the red pigment that transports oxygen.

Oxyhaemoglobin is the product that is formed when oxygen from the lungs combines with haemoglobin in the blood.

Affinity for means 'an attraction to' — it means that the haemoglobin attracts and holds the oxygen.

Oxyhaemoglobin dissociation curves

Haemoglobin has a high affinity for oxygen. The amount it takes up depends on the amount of oxygen in the surrounding tissues and is measured by its partial pressure of oxygen (pO_2) or oxygen tension.

At a low pO_2, haemoglobin does not readily take up oxygen molecules. The haem groups are hidden at the centre of the haemoglobin molecule, making it difficult for the oxygen molecules to associate with them. This difficulty in combining the first oxygen molecule accounts for the low saturation level of haemoglobin at low pO_2.

As the pO_2 rises, one oxygen molecule succeeds in associating with one of the haem groups. This causes a conformational change in the shape of the haemoglobin molecule, which allows more oxygen molecules to associate with the other three haem groups more easily. This accounts for the steepness of the curve as pO_2 rises (Figure 5.8).

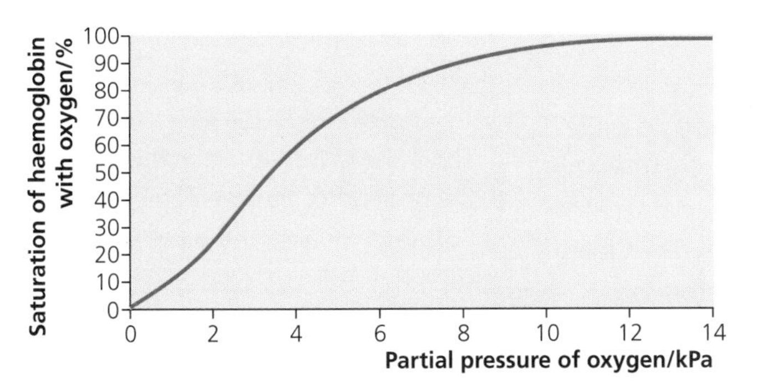

Figure 5.8 An adult oxyhaemoglobin dissociation curve

Now test yourself

8 Explain why a person using a spirometer containing air would soon feel tired and breathless.

Answer on p. 109

Tested

Releasing oxygen

In the body tissues, cells need oxygen for aerobic respiration, so oxyhaemoglobin releases the oxygen. This is called **dissociation**.

Dissociation is the release of oxygen from oxyhaemoglobin.

Fetal haemoglobin is a modified form of haemoglobin found in the mammalian fetus.

Fetal haemoglobin

The haemoglobin of a mammalian fetus has a higher affinity for oxygen than adult haemoglobin. Its haemoglobin must be able to take up oxygen from an environment that makes the adult haemoglobin release oxygen. In the placenta, the **fetal haemoglobin** must absorb oxygen. This reduces the oxygen tension near the blood, making the maternal haemoglobin release oxygen. Therefore, the oxyhaemoglobin dissociation curve for fetal haemoglobin is to the left of the curve for adult haemoglobin.

Typical mistake

Many candidates state that the fetal haemoglobin 'steals' or 'snatches' oxygen from the maternal haemoglobin, which is not accurate.

Transport of carbon dioxide

Revised

Carbon dioxide released from respiring tissues must be removed from the tissues and transported to the lungs. Carbon dioxide in the blood is transported in three ways:

- as hydrogen carbonate ions (HCO_3^-) in the plasma (85%)
- combined directly with haemoglobin to form a compound called carbaminohaemoglobin (10%)
- dissolved directly in the plasma (5%)

How are hydrogen carbonate ions formed?

As carbon dioxide diffuses into the blood, some of it diffuses into the red blood cells. Here, it combines with water to form a weak acid (carbonic acid). This reaction is catalysed by the enzyme carbonic anhydrase.

$$CO_2 + H_2O \rightarrow H_2CO_3$$

The carbonic acid dissociates to release hydrogen ions (H^+) and hydrogen carbonate ions (HCO_3^-).

$$H_2CO_3 \rightarrow H^+ + HCO_3^-$$

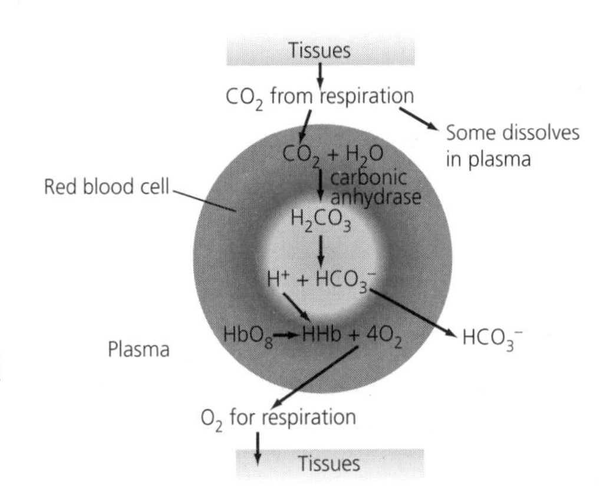

Figure 5.9 Most of the carbon dioxide diffuses into red blood cells and is converted to hydrogen carbonate ions by the action of carbonic anhydrase. Some carbon dioxide dissolves in the plasma

The hydrogen carbonate ions diffuse out of the red blood cells into the plasma. The charge inside the red blood cells is maintained by the movement of chloride ions (Cl^-) from the plasma into the red blood cells. This is called the chloride shift.

The hydrogen ions could cause the contents of the red blood cells to become acidic. To prevent this, the hydrogen ions are taken up by haemoglobin to produce haemoglobinic acid. The haemoglobin is acting as a buffer (a compound that can maintain a constant pH).

Examiner's tip

The detail of the creation and dissociation of carbonic acid is essential to gain all the marks.

The Bohr effect

The hydrogen ions released from the dissociation of carbonic acid must be taken up by the haemoglobin. However, the hydrogen ions cannot be taken up by oxyhaemoglobin. Effectively, the hydrogen ions compete for the space taken up by oxygen on the haemoglobin molecule. Therefore, when a lot of carbon dioxide is present, the hydrogen ions produced from carbonic acid displace the oxygen on the haemoglobin, so the oxyhaemoglobin releases more oxygen to the tissues. This is known as the **Bohr effect**. It results in the oxyhaemoglobin dissociation curve moving to the right.

The Bohr effect results in more oxygen being released where more carbon dioxide is produced from respiration. This is just what the muscles need to supply oxygen for aerobic respiration to continue.

The **Bohr effect** is the shift in the position of the haemoglobin dissociation curve in the presence of carbon dioxide.

Examiner's tip

It really does help to understand the role of hydrogen ions here — explaining this clearly is a high demand topic.

Exam practice

1 **(a)** Describe the initiation and coordination of one heart beat. [5]

(b) (i) In a condition known as superventricular tachycardia, the non-conducting fibres between the atria and the ventricles occasionally conduct the excitation wave. Suggest what effect this might have on the action of the heart. [2]

(ii) Suggest what effects this might have on the patient. [2]

2 Explain why the artery walls contain smooth muscle and elastic fibres. [4]

3 **(a)** State what creates the hydrostatic pressure at the arterial end of the capillary. [2]

(b) State two reasons why the pressure is much lower at the venous end of the capillary. [2]

4 Describe what is meant by an open circulatory system and explain why it is less efficient than a closed system. [3]

Answers and quick quiz 5 online

Online

Examiner's summary

By the end of this chapter you should be able to:

- Explain the need for transport systems in large and active organisms.
- Explain the difference between open/closed circulatory systems and single/double circulatory systems.
- Describe the internal and external structures of the mammalian heart.
- Describe the action of the heart and the cardiac cycle.
- Describe the structure of the blood vessels.
- Understand the differences between blood, tissue fluid and lymph.
- Understand how oxygen and carbon dioxide are transported.

6 Transport in plants

All living cells need a supply of oxygen and **nutrients** to survive. They also need to remove **waste** products such as oxygen so that they do not build up and become toxic. There are three main factors that affect the need for a transport system:

- size
- **surface area to volume ratio**
- level of activity

Multicellular plants are large organisms, so they need a transport system.

Typical mistake

Many candidates describe small organisms as having a large surface area rather than a large surface area to volume ratio, or they describe a large organism as having a small surface area.

Examiner's tip

Do not confuse surface area with surface area to volume ratio. It may help to always write SA/Vol rather than surface area to volume ratio.

Xylem and phloem

Distribution and function

Revised

There are two types of transport tissue in plants (Figure 6.1):

- **xylem** tissue moves water and minerals from the **roots** to the **leaves**
- **phloem** tissue moves **assimilates** up and down the plant from sources to sinks

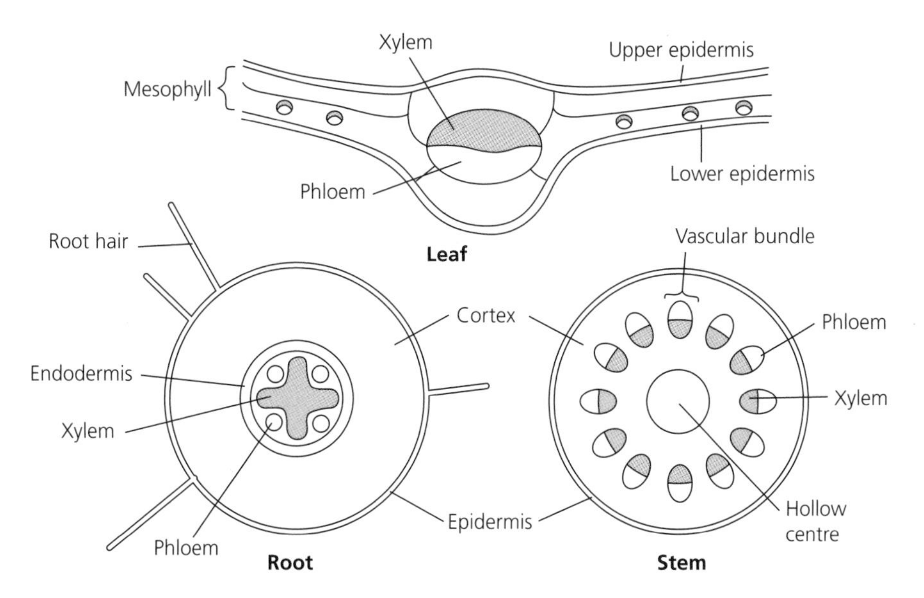

Figure 6.1 The distribution of xylem and phloem in cross-sections of leaf, root and stem

Revision activity

From memory, draw the position of the xylem and phloem in roots and stems.

Adaptations of xylem

Revised

The xylem is adapted to enable the free flow of water along the vessels. The cells have been killed by impregnation of the walls with **lignin**.

Xylem vessels have several adaptations (Figure 6.2):

- end walls are removed to form long tubes
- no cytoplasm or organelles are present
- cell walls are impregnated with lignin (lignified) to make the vessel wall waterproof and strengthen the vessel to support against collapse
- spiral, annular and reticulate thickening strengthens the wall to prevent collapse
- bordered pits between the vessels allow the movement of water between vessels

Examiner's tip

Make sure you say that *cell walls* have been lignified — do not say that the vessels or the xylem have been lignified. Also remember that this makes the walls waterproof — the xylem itself is not waterproof as there are pits to allow water in and out.

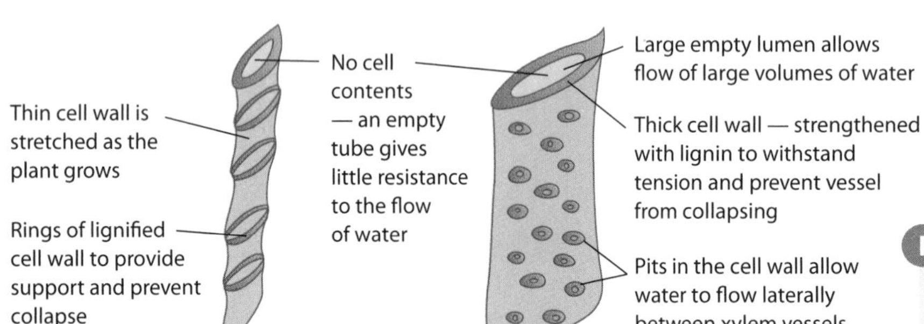

Figure 6.2 Two xylem vessel elements: (a) A narrow vessel thickened with rings (b) A wider vessel with pits to allow lateral movement of water

Revision activity

Draw a diagram of a xylem vessel and annotate it with the features that adapt it to transporting water.

Now test yourself

Tested

1 Explain why the xylem vessel walls are impregnated with lignin and why it is important to have pits in the walls.

Answer on p. 109

Adaptations of phloem

Revised

The phloem is adapted to transport assimilates actively by mass flow. There are two cell types involved:

- **sieve tube elements** — long sieve tubes that transport the assimilates
- **companion cells** — support cells that provide all the metabolic functions for the sieve tube elements and are involved in actively loading the sieve tubes

Phloem cells have several adaptations, as summarised in Table 6.1.

Table 6.1 Adaptations of phloem cells

Cell	Adaptation
Sieve tube elements	Form long tubes End walls are retained End walls contain many sieve pores, so they are called sieve plates Thin layer of cytoplasm Very few organelles and no nucleus
Companion cells	Closely associated with sieve tube elements Connected to sieve tube elements by many plasmodesmata (singular: plasmodesma) Dense cytoplasm with many mitochondria Large nucleus

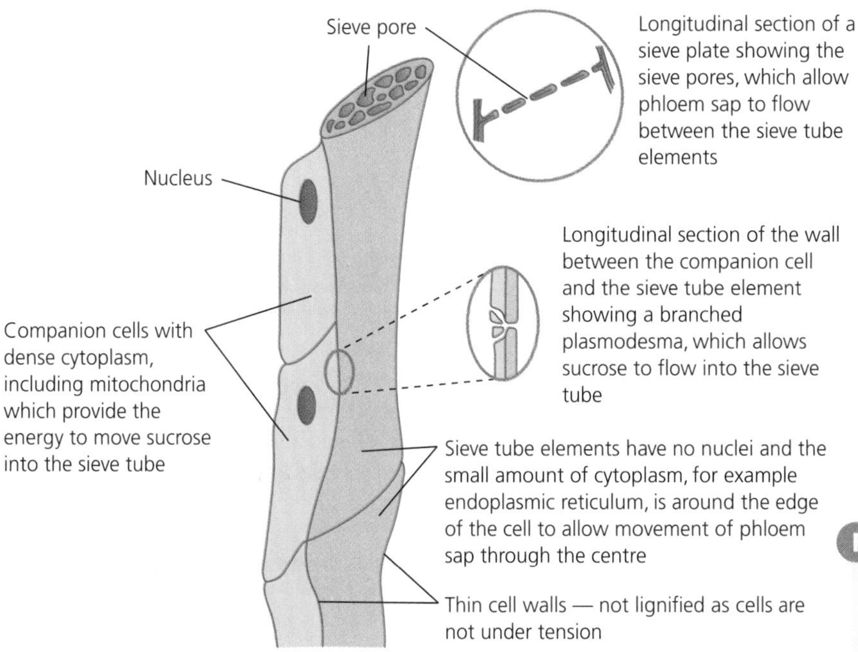

Figure 6.3 Phloem sieve tubes and companion cells, showing how they are adapted for their functions. Plasmodesmata are small tubes of cytoplasm that pass through the cell wall. They are lined by membrane continuous with the cell surface membrane

Revision activity

Draw a length of phloem tissue and annotate it with the features that adapt it to transporting assimilates.

Now test yourself Tested

2 Explain why the companion cells are essential.

Answer on p. 109

Transpiration

Loss of water vapour

Revised

Transpiration is the loss of water vapour from the upper parts of a plant, mainly the leaves. Some water will evaporate and diffuse through the leaf surface. However, most vapour is lost via the **stomata**. Transpiration involves three stages:

1 Water moves by osmosis from the xylem to the mesophyll cells in the leaf.

Transpiration is the loss of water vapour from the aerial parts of the plant.

2 Water evaporates from the surfaces of the spongy mesophyll cells into the air spaces inside the leaves.
3 Water vapour diffuses out of the leaf via the stomata.

Typical mistake

Many candidates state that water is lost from the leaf — it is water *vapour* that is lost.

The stomata open during the day to allow gaseous exchange — carbon dioxide enters the leaf and oxygen is released. This is to enable **photosynthesis** to occur. As the stomata are open, water vapour is lost. Transpiration is therefore a consequence of **gaseous exchange**.

Revision activity

Sketch a cross-section of a leaf and draw arrows to show the movement of water molecules. Label each arrow and explain what is happening to make the water move.

Water potential gradient

Revised

There must be a water potential gradient between the air spaces in the leaf and the surrounding air to make water vapour leave the leaf. The steeper this gradient, the more rapid the loss of water vapour (transpiration).

Factors that increase **transpiration rate** include:

- higher temperatures — this increases evaporation so there will be a higher water potential inside the leaf
- more wind — this blows water vapour away from the leaf, reducing the water potential in the surrounding air
- lower relative humidity — this increases the water potential gradient between the air inside the leaf and outside
- higher light intensity — this causes the stomata to open wider

Examiner's tip

To gain the marks, you must stress that higher temperatures increase transpiration, not simply that temperature increases transpiration.

Typical mistake

Many candidates confuse water potential and water potential gradient. A gradient must be between two points such as between the air inside the leaf and the air outside.

Typical mistake

Some candidates describe water leaving the leaf or being blown away from the leaf's surface. Remember that you must refer to water vapour.

Now test yourself

Tested

3 Explain why transpiration is quicker on a hot sunny day than on a cool cloudy day.

Answer on p. 109

Measuring transpiration rate

Revised

A bubble **potometer** (Figure 6.4) can be used to estimate transpiration rates. A potometer actually measures water uptake by the stem, but you can assume that water uptake equals water loss from the leaves in most cases. Care must be taken when setting up the potometer to ensure that there are no leaks and no air in the system, except the bubble used for measuring. Once the shoot has been allowed to acclimatise, the movement of the bubble along the capillary can be measured under different conditions. Transpiration rate is calculated by dividing the distance moved by a set time.

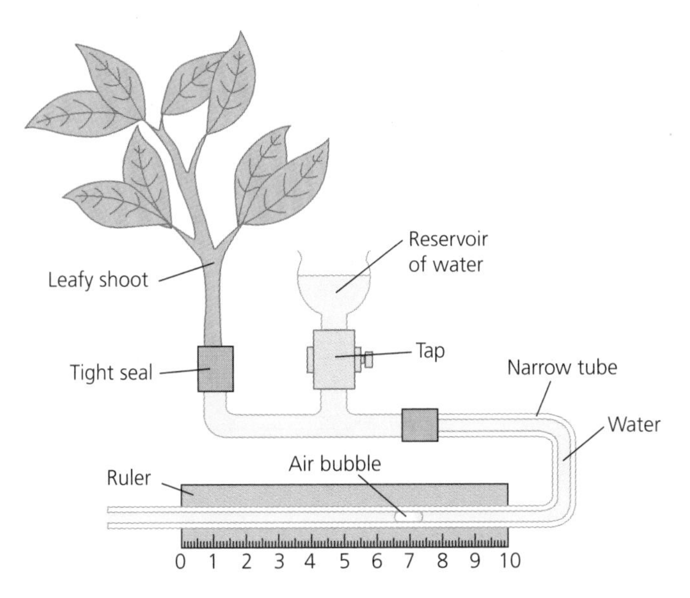

Figure 6.4 A typical potometer for measuring the rate of water uptake by leafy shoots

Revision activity

List the precautions you should take when setting up a bubble potometer.

Movement of water from cell to cell

Revised

The cell wall of a plant cell is permeable to water. The cell surface membrane is selectively permeable. As a result, **plant cells** can contain mineral ions in solution which reduces the **water potential** inside the cell. The more concentrated the mineral ions, the lower the water potential. Water moves by osmosis from a cell with a higher water potential to a cell with a lower water potential because water molecules move down their water potential gradient (Figure 6.5).

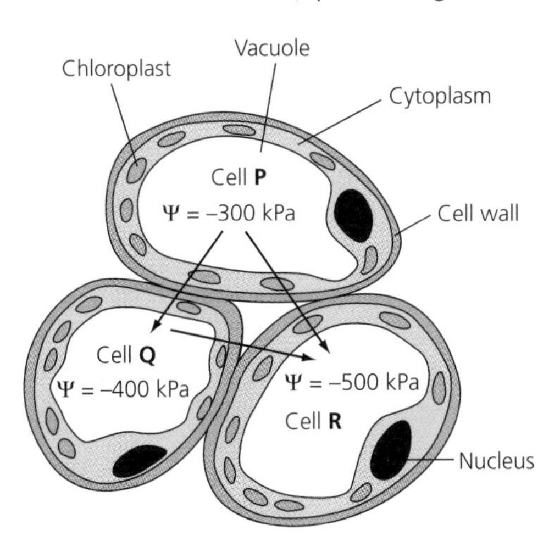

Figure 6.5 The arrows show the direction of water movement between three mesophyll cells in a leaf. Cell P has the highest water potential and the water potential of cell Q is higher than that of cell R

Typical mistake

Many students treat –1200 kPa as higher than –1100 kPa. Remember that the figures are negative. It may be best to describe one as more negative than the other.

Similarly, water can enter a cell from its **environment** if the water potential in the cell is lower than the water potential in the environment. This is how root hair cells absorb water from the **soil**.

Pathways

Once in the plant, water can move across a tissue such as the **root cortex** by different pathways (Figure 6.6):

- the **apoplast pathway** carries water between the cells through the cell walls — the water does not enter the cytoplasm or pass through cell surface membranes

- the **symplast pathway** takes water from cell to cell through the cytoplasm of each cell. Water often passes through plasmodesmata linking the cytoplasm of adjacent cells
- the **vacuolar pathway** carries water through the cytoplasm and vacuole of each cell

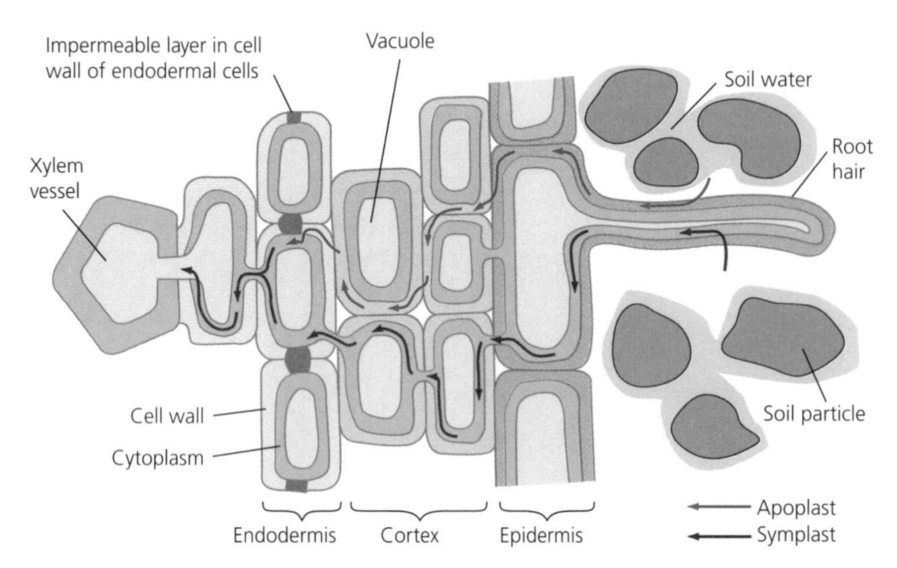

Figure 6.6 The arrows show the pathways taken by water as it moves from the soil, into a root hair, across the cortex, through the endodermis and into the xylem. Plasmodesmata are shown as the shaded areas passing through the cells walls, but they are not as big as shown here. Root hair cells provide a large surface area for the absorption of water

The transpiration stream

Water movement from the roots up to the leaves in the xylem is known as the **transpiration stream**. There are three mechanisms that move water up the **stem**:

- **root pressure**
- **adhesion** or **capillary action**
- transpiration pull

Root pressure and capillary action combined can only raise water by a few metres. Therefore, transpiration and the pull it creates are essential to move water all the way up a tall stem.

Transpiration stream is the movement of water from the roots to the leaves.

Root pressure is the pressure created by the action of the endodermis.

Adhesion is the attraction between water molecules and the walls of the xylem.

Examiner's tip

Remember that transpiration is the loss of water vapour from the leaf. Transpiration stream is the flow of water from the roots to the leaves to replace the water lost in transpiration.

Now test yourself Tested

4 Explain the difference between transpiration and transpiration stream.

Answer on p. 109

Root pressure is created by the action of the endodermis in the roots. The endodermis uses metabolic energy to pump mineral ions into the root medulla. This reduces the water potential in the medulla and xylem, making it more negative than in the cortex. Therefore, water moves across the endodermis into the medulla by osmosis.

Water cannot return to the cortex through the apoplast pathway as this is blocked by the **Casparian strip**. Therefore, pressure builds up in the cortex, which pushes the water up the xylem.

Adhesion is the attraction between the water molecules and the walls of the xylem vessel. It results in the water creeping up the xylem in a process called capillary action.

Loss of water vapour from the leaves must be replaced by water in the xylem. As water moves out of the xylem, it creates a pull on the column of water in the xylem. As water is cohesive (the molecules attract one another), the column of water is put under tension and pulled up the stem. This is known as the **cohesion tension theory**.

The **cohesion tension theory** accounts for the movement of water up the xylem.

Typical mistake

Some candidates state that the water moves up the xylem by cohesion tension. This is incorrect — water moves up the xylem as a result of tension created by the loss of water in the leaves, which draws the whole column of water up the xylem due to the cohesion between the water molecules.

Movement of water up the stem

Water vapour lost from the leaves reduces the hydrostatic pressure of water in the xylem at the top of the plant. This creates a hydrostatic pressure gradient between the high pressure created in the roots and the low pressure in the leaves. Water moves up the stem from high to low pressure.

Typical mistake

Candidates sometimes confuse the movement of water due to hydrostatic pressure differences with movement between cells caused by a water potential gradient. Water potential gradients cannot move water all the way up the xylem.

Revision activity

Draw a mind map to link together the ideas about how water moves up the xylem.

Xerophytes

Revised

Xerophytes are plants that are **adapted** to living in dry (or arid) places. These adaptations help them to reduce loss of water vapour:

- thick waxy cuticle on the leaves
- smaller leaf area
- stomata in pits
- hairy leaves
- rolled leaves

Revision activity

Sketch a leaf from a xerophyte and annotate it with all the features that help the plant conserve water. For each feature, explain how it reduces water loss using the terms *water potential* and *water potential gradient*.

Translocation

Translocation is the movement of assimilates (mostly **sucrose**) around the plant. It occurs in the sieve tubes, but the companion cells are important in actively loading assimilates into the sieve tubes.

Translocation is the movement of assimilates around the plant.

Sources and sinks

Revised

A **source** is a part of the plant that has a supply of assimilates that are loaded into the phloem. This could be:

- a leaf that has made sucrose from the products of photosynthesis during the spring and summer
- a root that has stored starch and can convert this to sucrose, which happens particularly in spring
- any other storage organ where the plant has stored starch

A **sink** is a part of the plant that removes sucrose from the phloem and uses or stores it. This could be:

- the buds or stem tips where growth occurs and energy is needed
- the leaves in spring as they grow and unfold
- the roots in summer and autumn when the plant is storing sugars as starch
- any other organ where the plant may store starch

The sucrose can be used in respiration to provide energy or as a building block to produce larger molecules for growth.

A **source** is a tissue or organ that supplies assimilates to the phloem.

A **sink** is a tissue or organ that removes assimilates from the phloem and uses them.

Now test yourself

5 Explain how a leaf can be a source or a sink.

Answer on p. 109

Tested

The mechanism of translocation

Revised

Translocation is achieved by mass flow. It is caused by creating a high hydrostatic pressure at the source and a lower hydrostatic pressure at the sink. The fluid in the phloem sieve tube then moves from high to low pressure, i.e. down its pressure gradient. The evidence for and against the mechanism of translocation is given in Table 6.2.

Creating high pressure at the source

This process is called **active loading**. Sucrose is moved into the sieve tube by a complex process involving active transport:

1. Hydrogen ions are pumped actively out of the companion cells, which use ATP as a source of energy.
2. The hydrogen ions diffuse back into the companion cells through special co-transporter proteins carrying sucrose molecules into the companion cells.
3. The sucrose builds up in the companion cells and diffuses into the sieve tube through the many plasmodesmata.
4. The water potential in the sieve tube is reduced.
5. Water flows into the sieve tube by osmosis, increasing the pressure.

Revision activity

Draw a flow chart to show how the sieve tube elements are actively loaded.

Examiner's tip

The details of the mechanism of translocation are described here in some detail, which is needed to achieve full marks.

Creating lower pressure at the sink

As sucrose is used in respiration or converted to starch in the cells of the sink, the concentration decreases. This creates a concentration gradient between the sieve tubes and the cells in the sink. Sucrose diffuses out of the sieve tubes into the cells and the water potential in the sieve tube increases. Water then moves out of the sieve tube by osmosis.

Table 6.2 Evidence for and against the mechanism of translocation

Evidence that the phloem is involved	Radioactive-labelled carbon supplied to the leaves as carbon dioxide in the air appears in the sieve tubes soon after being supplied Ringing the plant (removing the bark and phloem in a ring around the stem) kills the plant by stopping translocation
Requirement for energy	Metabolic poisons stop translocation Rate of flow is too high for simple diffusion Companion cells contain many mitochondria
Evidence for translocation	pH of companion cells is lower than surrounding cells Sucrose concentrations are always higher in the source than in the sink There are many plasmodesmata between companion cells and sieve tube elements
Evidence against translocation	Different solutes move at different rates in the phloem Sucrose moves to all parts of the plant at the same rate instead of being moved more quickly to areas with a greater need Some electron micrographs show sieve plate pores blocked with protein It is not clear what the role of the sieve plates is

Exam practice

1 **(a)** Define the term *transpiration*. [2]

(b) Explain why transpiration is an inevitable result of photosynthesis. [2]

(c) Many xerophytes have a thick waxy cuticle and roll their leaves. Explain how these features reduce transpiration. In your answer, you should use appropriate technical terms, spelt correctly. [5]

2 **(a)** Define the terms *source* and *sink*. [3]

(b) Name two possible sources. [2]

(c) Explain how a leaf can be a sink and a source at different times. [2]

3 **(a)** Describe the features of the sieve tubes that help mass flow to occur. [3]

(b) Describe and explain how the companion cells are specialised to their role in loading assimilates into the sieve tubes. [3]

Answers and quick quiz 6 online

Online

Examiner's summary

By the end of this chapter you should be able to:

- ✔ Explain why large plants need a transport system.
- ✔ Describe the distribution of xylem and phloem.
- ✔ Describe the structure of xylem and phloem and how they are adapted.
- ✔ Define transpiration, describe the factors that affect the rate of transpiration and explain how to measure the rate using a potometer.
- ✔ Explain, using the term *water potential*, how water moves between cells and across plant tissues.
- ✔ Describe and explain how water moves up the xylem.
- ✔ Describe and explain how assimilates are moved in the phloem.

7 Biological molecules

Water

Hydrogen bonds

Revised

Water is a simple molecule that can form **hydrogen bonds** between its molecules. This gives water some important properties.

A **hydrogen bond** is a weak bond in which a hydrogen atom in one molecule is attracted to an electronegative atom (oxygen) in the same or a different molecule.

Examiner's tip

In any response, ensure you relate the properties of water to the hydrogen bonding.

A hydrogen bond is a force of attraction between the oxygen atom in one water molecule and the hydrogen atoms in the same or another water molecule. This attraction occurs because each water molecule is polar, which means there is an uneven distribution of charge. Oxygen atoms attract electrons more strongly than hydrogen atoms. Therefore, the electrons in a water molecule are pulled towards the oxygen atom, which gives the oxygen a more negative charge. However, the electrons are not pulled completely on to the oxygen atom, so the charge on the oxygen is not completely negative. This charge is shown as delta negative ($\delta-$). The hydrogen in the molecule is left with a delta positive ($\delta+$) charge. It is these opposite charges that come into force when producing hydrogen bonds (Figure 7.1a).

When more molecules are present, as in the case of liquid water, more bonds are possible because the oxygen atom of one water molecule has two lone pairs of electrons, each of which can form a hydrogen bond with a hydrogen atom on another water molecule. This can repeat so that every water molecule is bonded with up to four other molecules (Figure 7.1b).

(a)

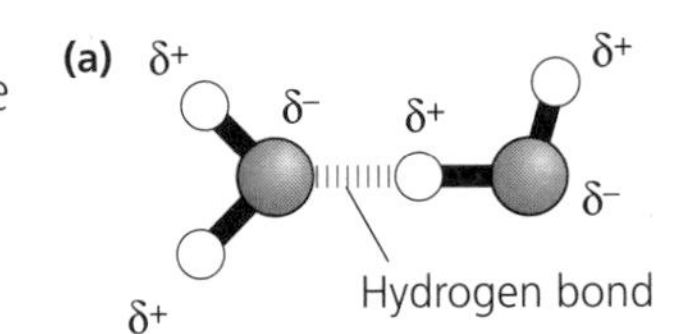

(b)

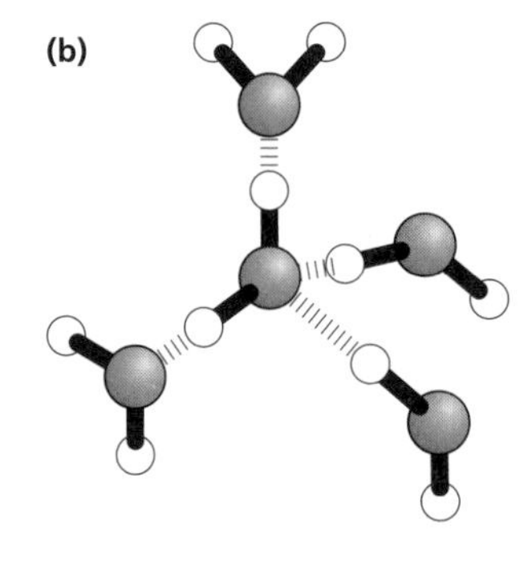

Figure 7.1 (a) Two water molecules with a hydrogen bond between them (b) A cluster of water molecules held together by hydrogen bonds

Examiner's tip

Always draw hydrogen bonds as dashed lines.

How do hydrogen bonds affect the properties of water?

Hydrogen bonds cause cohesion. Between 0°C and 100°C, hydrogen bonds hold water molecules together loosely — they are held together, but they can move past one another and the water remains a liquid.

In order to evaporate, the hydrogen bonds must be broken, allowing the molecules to separate and form a gas (water vapour). This takes a lot of energy, so water remains a liquid up to 100°C.

At lower temperatures the molecules have less kinetic (movement) energy and move about less. With less movement, more hydrogen bonds can form and at 0°C enough hydrogen bonds have formed to hold the water molecules in a stationary position, forming ice. The water molecules are now held in a rigid lattice, which keeps the molecules further apart. Therefore, ice floats as it is less dense than water.

Now test yourself

1 Explain what is meant by a $\delta+$ charge and how it is created.

Answer on p. 109

Tested

Typical mistake

Many candidates fail to make the link between the action of the hydrogen bonds and the properties of water.

Water properties and living organisms

Revised

The following **properties of water** are important for the survival of many **living organisms**:

- Thermal stability — water has a high specific heat capacity, which means that a lot of energy is needed to warm it up. Therefore, a body of water maintains a fairly constant temperature, which is essential for life to survive.
- Freezing — ice is less dense than water so it floats, which insulates the water and prevents it freezing completely. Living things can survive below the ice.
- Evaporation — a lot of energy is needed to cause evaporation, which is used to cool the surface of living things. This energy is known as latent heat. Water has a high specific latent heat capacity.
- At most common temperatures water is a liquid — it can flow and transport materials in living things.
- **Cohesion** — the attraction of water molecules produces surface tension, which creates a habitat on the surface. It also enables continuous columns of water to be pulled up the xylem.
- Solvent — as the molecules are polar, water can dissolve a wide range of substances.
- As a reactant — water molecules are used in a wide range of reactions such as hydrolysis and photosynthesis.
- Incompressibility — water cannot be compressed into a smaller volume. This means it can be pressurised and pumped in transport systems or used for support in hydrostatic skeletons.

Revision activity

Draw a mind map with water in the centre to show how its properties are essential for living things.

Cohesion is the weak attraction between two similar molecules.

Now test yourself

2 Explain the difference between specific heat capacity and latent heat capacity.
3 What roles do specific heat capacity and latent heat capacity play in the survival of living things?

Answers on p. 109

Tested

Amino acids and proteins

Amino acids

Revised

Proteins are made up of long chains of **amino acids**. There are 20 different amino acids used in proteins, but all have the same basic structure (Figure 7.2). The *R* group is the only part that differs between different amino acids.

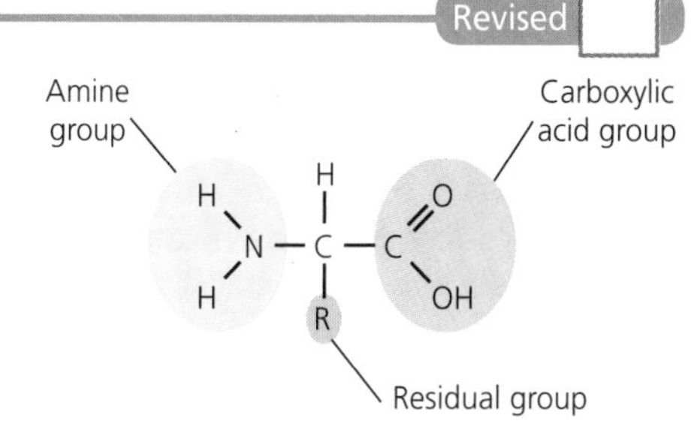

Figure 7.2 The generalised structure of an amino acid

Proteins

Revised

All proteins consist of unbranched chains of amino acids, which are held together by **peptide bonds**. These bonds are formed by **condensation** and occur between the amine group of one amino acid and the carboxylic acid group of another.

A **peptide bond** is the bond between two amino acids.

Condensation is a reaction that involves the release of water molecules.

Two amino acids together make a **dipeptide** (Figure 7.3). Many amino acids in a chain form a **polypeptide**.

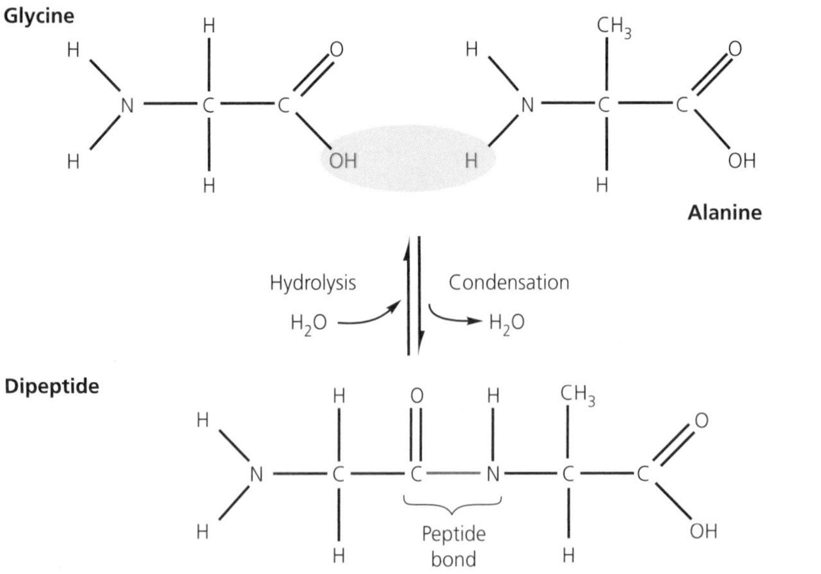

A **dipeptide** is formed when two amino acids are joined together by a peptide bond.

A **polypeptide** is a chain of many amino acids joined together by peptide bonds.

Figure 7.3 The formation and breakage of a peptide bond between glycine and alanine

Now test yourself

4 Explain why proteins are unbranched chains.

Answer on p. 109

Tested

Protein structure

There are four levels of protein structure.

A **primary structure** is a sequence of amino acids in a chain, held together by peptide bonds.

A **secondary structure** is formed when the chain of amino acids becomes folded and coiled. Two shapes are formed:

- alpha (α) helix — shaped like a coil spring
- beta (β) sheets — pleated like a folded sheet of paper

Hydrogen bonds hold the folds and coils in place.

A **tertiary structure** is formed when the coiled and pleated chains can be folded further to produce the final three-dimensional shape of the molecule. These final folds and coils are caused by the interactions between the *R* groups on the amino acids, which interact to form a range of bonds that hold the three-dimensional shape.

These bonds include:

- hydrogen bonds between polar *R* groups
- ionic bonds between *R* groups with opposite charges
- covalent disulfide bonds between two sulphur-containing cysteine amino acid residues to form disulfide bridges

Some *R* groups are **hydrophobic** and twist away from water into the centre of the molecule. Others are **hydrophilic** and twist outwards so that they are on the outside of the molecule.

Typical mistake

Some candidates suggest that a protein shows just one of the levels of structure or that the labelling applies to the nutritional value of the protein. This is not the case — all proteins show a primary, secondary and tertiary level of structure, and some show a quaternary level of structure.

Examiner's tip

A typical question might ask you to relate the structure of a protein to its properties — this could be answered in the form of a table.

Hydrophobic means water-hating, repelled by water.

Hydrophilic means water-loving, attracted to water.

Many proteins are just one polypeptide chain that has been coiled and folded. However, some proteins consist of more than one polypeptide chain. For example, haemoglobin has four polypeptide chains and collagen has three. These multi subunit proteins make up the **quaternary structure**.

Now test yourself

5 Explain the difference between a polypeptide and a protein.

Answer on p. 109

Tested

Collagen

Revised

Collagen is a **fibrous protein**. Its structure provides strength in places such as in the walls of arteries and in tendons that attach muscle to bone. It has three chains of polypeptides that wind around one another. Each chain consists of about 1000 amino acids arranged in a helix.

There are only a few different amino acids and every third amino acid is glycine, which has a small *R* group. This allows the polypeptides to coil tightly to provide greater strength. The three polypeptide chains are held firmly together by many hydrogen bonds.

Many collagen molecules are joined together by covalent bonds. The molecules are arranged to overlap each other so that there are no weak points where they join together.

Haemoglobin

Revised

Haemoglobin is a **globular protein**. It is used to transport oxygen in the form of oxyhaemoglobin.

Haemoglobin contains four polypeptide chains called subunits (two alpha-chains and two beta-chains). Each subunit has a non-protein prosthetic group attached called a haem group, which contains a single iron ion (Fe^{2+}). One oxygen molecule can attach to each haem group, so a haemoglobin molecule can carry four oxygen molecules.

Revision activity

Make a model (or draw a diagram) of a protein using coiled wire or a slinky spring and folded paper. Label and annotate the model to show examples of the different types of bonding.

Now test yourself

Tested

6 Explain why proteins that are metabolically active such as enzymes or hormones are globular proteins.
7 Explain why globular proteins are more affected by temperature change than fibrous proteins.

Answers on p.109

Carbohydrates

Carbohydrates consist of carbon, oxygen and hydrogen. They are based on simple subunits called **monosaccharides**. **Disaccharides** and **polysaccharides** are complex carbohydrates made up of chains of monosaccharide subunits.

Alpha-glucose and beta-glucose

Revised

The structures of **alpha-glucose** and **beta-glucose** are shown in Figure 7.4.

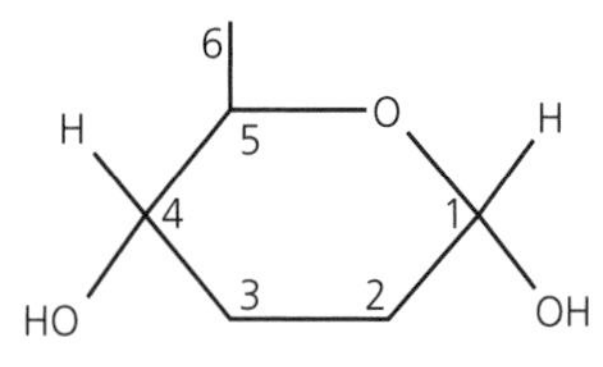

α-glucose (note that the –H is above the –OH on carbon atom 1)

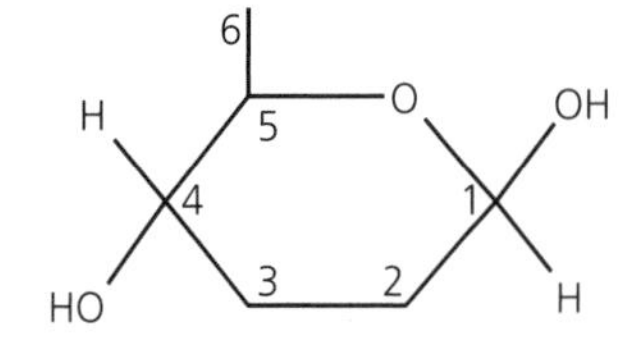

β-glucose (note that the –OH is above the –H on carbon atom 1)

Figure 7.4 Alpha-glucose and beta-glucose

The difference between alpha-glucose and beta-glucose is simply the position of the H and OH groups on the first carbon atom. This difference is not great, but it has a large effect on the way they bond together and the polysaccharides produced.

Disaccharides

Revised

Disaccharides are formed by a condensation reaction between two monosaccharides. The most common bond is between the carbon 1 of one monosaccharide and the carbon 4 of another. This forms a 1,4 **glycosidic bond** and releases water.

Glucose and fructose combine to form sucrose (Figure 7.5), whereas glucose and glucose combine to form **maltose**. Disaccharides can be converted back to monosccharides by **hydrolysis**.

A **glycosidic bond** is the link between two monosaccharides in a polysaccharide.

Hydrolysis is splitting a large molecule into two smaller molecules by adding water.

Glucose + **Fructose** → Condensation reaction → H_2O + **Sucrose**

Figure 7.5 Forming a 1,2 glycosidic bond between glucose and fructose to form sucrose

Polysaccharides

Revised

Starch

Starch is a combination of two molecules: **amylose** and amylopectin. Amylose consists of a long unbranched chain of alpha-glucose subunits. These subunits are joined by 1,4 glycosidic bonds. The chain of subunits coils up.

Examiner's tip

Complex carbohydrates are not that complex — they are simply long chains of one or two subunits. Each chain is called a polymer.

The hydroxyl groups on carbon 2 of each subunit is hidden inside the coil. This makes the molecule less soluble.

Amylose is used for the storage of glucose subunits and energy in plant cells. The molecule is compact — it takes little space in the cell. It is insoluble, which means the molecules do not affect the water potential of the cells.

Glucose subunits can be removed easily from each end of the molecule. These can be used as building blocks to build other substances or as a substrate in respiration to release stored energy.

Glycogen

Glycogen is similar to amylose. Some of the glucose subunits have a 1,6 glycosidic bond as well as the 1,4 glycosidic bond. This means that the molecule is branched.

Glycogen is used for the storage of glucose subunits and energy in animal cells. It has the same advantages as amylose but the branched molecule means that glucose subunits can be removed more quickly, allowing rapid bursts of energy supply.

Cellulose

Cellulose consists of a long unbranched chain of beta-glucose subunits. These subunits are joined by a 1,4 glycosidic bond. The chain of beta-glucose subunits form a straight chain.

The hydroxyl groups on carbon 2 of each subunit are exposed, allowing hydrogen bonds to form between adjacent cellulose molecules. Some 60–70 molecules bind together to form a cellulose microfibril and many microfibrils join together to form macrofibrils.

Cellulose is strong and completely insoluble. It is used in plant cell walls and provides enough strength to support the whole plant.

Examiner's tip

You only need to know about amylose, but you should be able to apply that knowledge to other complex carbohydrates.

Typical mistake

Many candidates are mistaken in thinking that cellulose is a protein, not a carbohydrate.

Examiner's tip

The structure of complex carbohydrates lends itself to a question in which the examiner asks you to relate the structure of a molecule to its function. This is perhaps most easily done as a table.

Now test yourself

8 List the reasons why amylose is a good storage product.
9 Explain why cellulose is insoluble.

Answers on p. 110

Tested

Revision activity

- Draw a table to compare the structures and properties of amylose, glycogen and cellulose.
- Draw a table to compare the structures and properties of cellulose and collagen.

Lipids

Lipids are not polymers like proteins and complex carbohydrates. They are a large group of compounds that includes triglycerides, phospholipids and steroids. Lipids are insoluble in water.

Triglycerides

Revised

A **triglyceride** is a combination of one glycerol molecule and three fatty acid chains (Figure 7.6). The fatty acids are attached to the glycerol by a condensation reaction. The bonds are called ester bonds.

A **triglyceride** contains one glycerol molecule and three fatty acid chains.

Triglycerides are rich in energy and used to store excess energy. When required, the molecules can be broken down in aerobic respiration to release this energy. Water is also released, which can be useful for animals that live in dry environments — hence camels store fat in their humps. The stores can be held under the skin and around major organs, which has the benefit of protecting the major organs from physical shock.

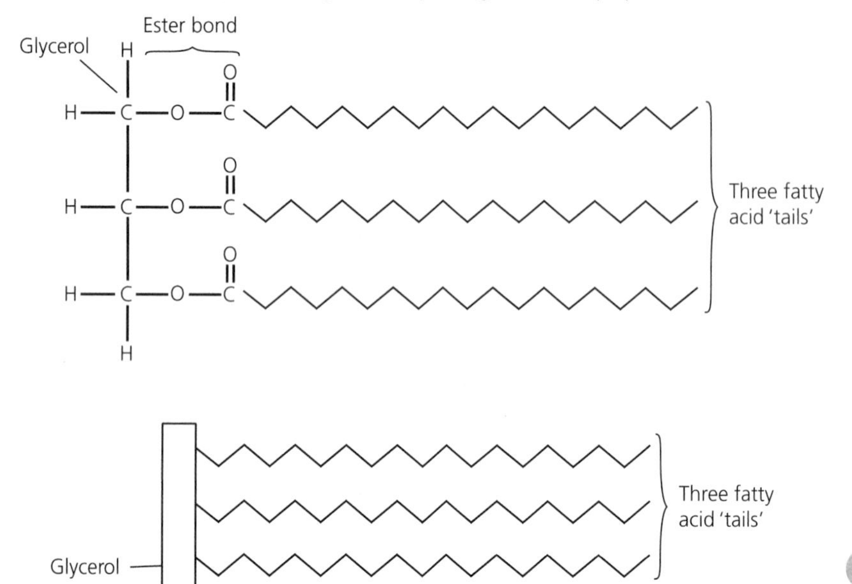

Figure 7.6 Two ways of showing a triglyceride

Triglyerides are also good insulators, and are used to insulate animals that live in cold environments such as polar bears and aquatic mammals such as whales. They also provide buoyancy for these mammals.

Now test yourself

10 Explain why lipids are good storage compounds.

Answer on p. 110

Tested

Phospholipids

Revised

Phospholipids are similar to triglycerides, but one of the fatty acid chains is replaced by a phosphate group. The two remaining fatty acid 'tails' are insoluble in water and are called hydrophobic.

The phosphate group is complex and includes choline, which is water soluble. This alters the characteristics of the molecule. This group makes the 'head' end of the phospholipid able to mix with water — it is hydrophilic.

Phospholipids form bilayers with the hydrophobic 'tails' in the centre and the hydrophilic 'heads' pointing outwards to interact with the surrounding aqueous solution. This is the basis of all cell membranes.

A **phospholipid** contains one glycerol molecule, a phosphate group and two fatty acid chains.

Now test yourself

11 Explain why phospholipids are used in membranes.

Answer on p. 110

Tested

Cholesterol

Revised

Cholesterol is similar to a phospholipid in that it has hydrophilic and hydrophobic regions. It is insoluble in water, but can mix with phospholipids. It has a number of functions in the body, such as:

- stabilising cell membranes, making them less fluid
- used in the manufacture of steroid hormones such as oestrogen
- waterproofing the skin
- used in the skin to manufacture vitamin D

Revision activity

Write an exam-style question worth 10 marks about the roles of lipids in living things. Ensure that at least 5 marks test assessment objective 2 (use of your knowledge in an unfamiliar context). Write out a mark scheme.

Testing for biological molecules

Specific tests

Revised

The presence of each biological molecule can be detected by a specific test. Table 7.1 summarises the procedures to be followed.

Typical mistake

This is straightforward recall, yet many candidates don't seem to learn the details. You must learn all these tests.

Table 7.1 Chemical tests to identify the presence of biological molecules

Molecule tested for	Test name	Details of test	Positive result
Protein	Biuret test	Dissolve in water. Add biuret A and biuret B	Colour change from blue to mauve or purple
Reducing sugar (e.g. glucose)	Benedict's test	Dissolve in water. Add Benedict's reagent. Heat at 80–90°C for 2 minutes	A precipitate forms. Colour change from blue to brick red
Non-reducing sugar (e.g. sucrose)	Benedict's test (this test converts non-reducing sugars to reducing sugars and then tests the reducing sugars)	Test the substance for reducing sugars to ensure that none are present, then dissolve in water. Add a few drops of dilute hydrochloric acid and boil for 2 minutes. Neutralise the solution by adding a few drops of dilute sodium hydroxide. Add Benedict's reagent and reheat for 2 minutes	A precipitate forms and there is a colour change from blue to brick red
Starch (amylose)	Iodine solution test	Dissolve in water. Add iodine solution	Colour change to deep blue/black
Fat (lipids)	Emulsion test	Dissolve in alcohol. Filter. Add water to the filtrate	When water is added to the clear filtrate, it will turn cloudy or milky

Making the Benedict's test quantitative

Revised

When testing for reducing sugars such as **glucose**, the colour change may not be complete: the colour may show a change from blue to green or yellow before going orange or red. If only a small amount of sugar is present, it will not react with all the Benedict's reagent — leaving some of it blue. This is what causes an incomplete colour change.

The Benedict's test can be made quantitative (i.e. you can determine the **concentration** of the reducing sugars) by ensuring there is excess Benedict's reagent. Create a range of colours using known concentrations of reducing sugars to create a set of colour standards.

To make the measurement more precise, the standards and the sample should be placed in a centrifuge and spun for 2 minutes. This will deposit the coloured precipitate at the bottom of the tube, leaving a blue solution of unused Benedict's reagent. The more concentrated the reducing sugar, the less blue colour will be left. The intensity of the blue colour in the solution can be measured using a **colorimeter**. Plot a graph of absorbance against concentration using the standard solutions. Use your graph to determine the concentration of the unknown solution by reading from the absorbance measurement across to the concentration.

Now test yourself

12 Explain why carrying out the Benedict's test on a solution of glucose at low concentration will leave the solution blue or green in colour, but if the glucose is at high concentration the solution will have no sign of blue colouration.

Answer on p. 110

Tested

Exam practice

1 **(a)** List three functions of carbohydrates in living things. [3]

(b) Describe and explain how the structure of starch (amylose) makes it suitable as a storage compound. In your answer you should make clear how the structure relates to the function. [6]

(c) Complete the following table to compare glycogen and cellulose. [5]

Structural feature	Glycogen	Cellulose
Sugar(s) present		
Bonds present		
Branched or unbranched		
Coiled or straight		
Forms cross-links with other molecules		

2 **(a) (i)** Explain what is meant by the term *primary structure of a protein.* [1]

(ii) Explain what is meant by the term *secondary structure of a protein.* [3]

(b) Describe the bonds involved in holding the tertiary structure of a protein. [4]

3 **(a)** Explain why hydrogen bonds form between water molecules. [3]

(b) Explain the role of hydrogen bonds in making water a suitable medium for transport in living things. [10]

4 A student carried out a test to determine if a certain substance was present in the seeds of a small plant. He crushed the seeds and added some alcohol. He then filtered the solution before adding water to the filtrate. The resulting solution turned milky.

(a) State what substance was tested for. [1]

(b) Why was it necessary to filter the solution? [2]

(c) Suggest how the student could make this test quantitative. [6]

Answers and quick quiz 7 online

Online

Examiner's summary

By the end of this chapter you should be able to:

- ✔ Describe how hydrogen bonding occurs.
- ✔ Relate the properties of water to its roles in living things.
- ✔ Describe the structure of amino acids and the formation of peptide bonds.
- ✔ Explain the four levels of protein structure with reference to the types of bonds involved at each level.
- ✔ Describe and compare the structures of haemoglobin and collagen as examples of a globular protein and a fibrous protein.
- ✔ Describe the structure of alpha-glucose and beta-glucose and the formation of glycosidic bonds.
- ✔ Describe and compare the molecular structures of amylose, glycogen and cellulose and explain how the structure of each molecule relates to its function.
- ✔ Describe and compare the structures of a triglyceride and a phospholipid and explain how the structure of each molecule relates to its function.
- ✔ Describe the chemical tests for protein, reducing sugars, non-reducing sugars, starch and lipids.
- ✔ Describe how the concentration of glucose can be determined using colorimetry.

8 Nucleic acids

DNA and RNA

Deoxyribonucleic acid (DNA)

Revised

Relating structure to function

The function of **deoxyribonucleic acid (DNA)** is to act as the genetic code. In order to do this, the DNA molecule must have three properties:

- the ability to carry coded information
- the ability to replicate precisely
- stability

DNA structure

DNA is a **polynucleotide**. This means it is made up of a long chain of units called **nucleotides**. It is usually **double stranded**.

Nucleotides

A nucleotide (Figure 8.1) has three components:

- a phosphate group
- a sugar (deoxyribose)
- an organic **base**: **adenine** (A), **thymine** (T), **cytosine** (C) or **guanine** (G)

Polynucleotides

A polynucleotide is formed when nucleotides bind together in a long chain. The bonds are formed between the sugar of one nucleotide and the phosphate group of another, making a sugar–phosphate 'backbone'. This leaves the organic base of each nucleotide sticking out to the side of the chain.

Forming a DNA molecule

In forming a DNA molecule, each type of base on one strand forms **hydrogen bonds** with a specific type of base on the other strand. This is called **complementary base pairing**:

- adenine is paired by two hydrogen bonds with thymine
- cytosine is paired by three hydrogen bonds with guanine

Adenine and guanine are both purine bases, whereas thymine and cytosine are both pyrimidine bases. A purine base always pairs with a pyrimidine base. This means that the two polynucleotide strands will run parallel. As the strands lie in opposite directions, we call this **antiparallel** (Figure 8.2).

Once the two single strands of polynucleotide are joined together to make a double strand, the whole molecule twists to form a **double helix** (Figure 8.3).

Now test yourself

1 Explain why DNA must be stable yet able to replicate precisely.

Answer on p. 110

Tested

A **nucleotide** is a combination of a phosphate, a pentose sugar and an organic base.

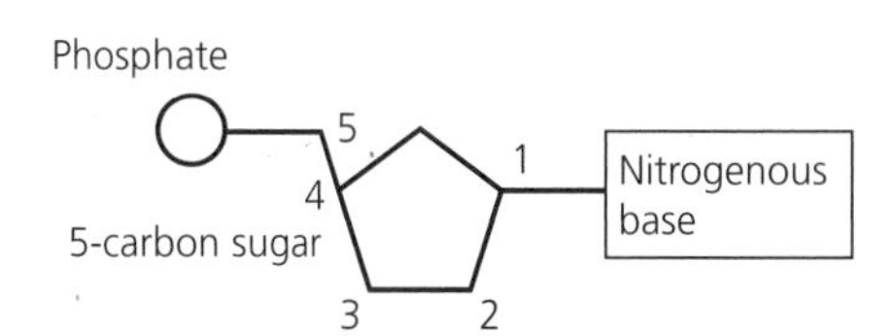

Figure 8.1 All nucleotides have this structure

Typical mistake

Many candidates don't recall that adenine is paired to thymine with two hydrogen bonds, whereas cytosine bonds to guanine with three. This is why adenine cannot bind to cytosine and thymine cannot bind to guanine.

Examiner's tip

The scientific research leading to the discovery of the structure of DNA is well recorded. It is an area where the examiner could test How Science Works.

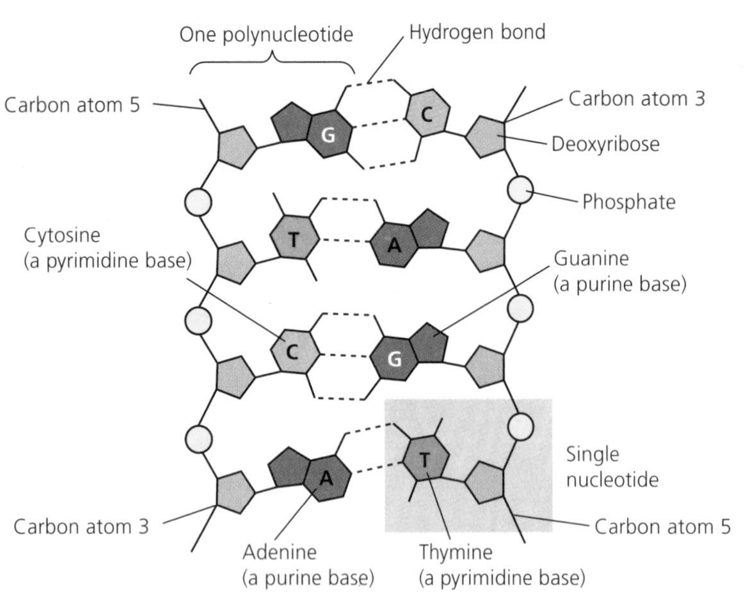

Figure 8.2 The molecular structure of DNA, showing two antiparallel polynucleotides

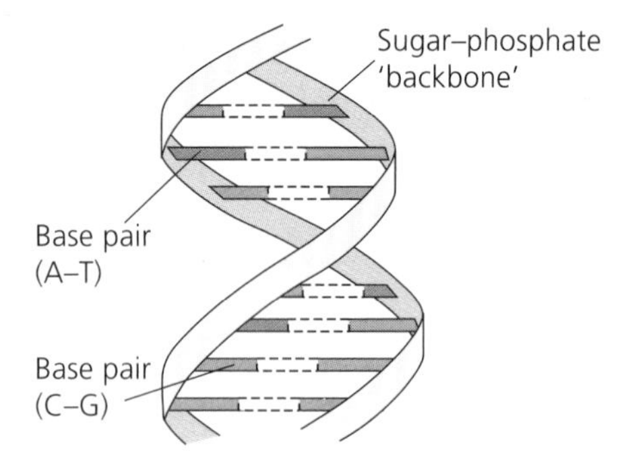

Figure 8.3 The two strands are twisted around each other to form the DNA double helix (this shows a small part of the double helix of DNA)

DNA replication

DNA **replication** occurs in all living organisms in order to copy their DNA for biological inheritance. Precise replication can take place because of the double-stranded structure of the DNA molecule. The process, known as **semi-conservative** replication, is as follows:

1. One double-stranded molecule untwists and the hydrogen bonds between the base pairs break.
2. The two polynucleotide chains separate, exposing the bases.
3. Each strand is then used as a template for a new strand.
4. New nucleotides pair to the exposed bases on both strands, using their complementary shapes to pair correctly.
5. The two new strands of nucleotides bond together to form the second half of the DNA molecule.
6. The enzyme **DNA polymerase** joins all the bases together. It also checks that the pairing is correct.
7. The new molecule then twists to form a double helix.

Replication means to make an identical copy.

Examiner's tip

The scientific research carried out by Meselson and Stahl to show that replication is semi-conservative is a classic example of investigative work and may be tested under How Science Works.

Revision activity

Draw a flow diagram to show the sequence of events in DNA replication.

Now test yourself

2 Explain why it is important that each new double helix is identical to the original.

Answer on p. 110

Tested

Ribonucleic acid (RNA)

Revised

Ribonucleic acid (RNA) is also a polynucleotide. It is usually **single stranded** and smaller than DNA.

The sugar involved is ribose sugar, not deoxyribose. The bases used include adenine, cytosine and guanine, but the thymine is replaced by uracil.

Typical mistake

Candidates sometimes forget that the sugar in RNA is ribose not deoxyribose, as found in DNA.

There are three types of RNA:

- messenger RNA (mRNA), which carries the code held in the genes to the ribosomes where the code is used to manufacture proteins
- transfer RNA (tRNA), which transports amino acids to the ribosomes
- ribosomal RNA (rRNA), which makes up the ribosome

Examiner's tip

Make sure that you can remember the key differences between DNA and RNA.

Revision activity

Draw a table to compare the structure of DNA with that of RNA.

The genetic code and protein synthesis

The genetic code

Revised

The **genetic code** is the set of rules by which information encoded in genetic material is translated into proteins (amino acid sequences) by living cells. DNA consists of a long **sequence** of DNA nucleotides specific to the species. The genetic code, which is a chain of bases, is carried in the sequence.

Three consecutive bases (called a triplet) code for one amino acid. A particular series of triplets will code for one sequence of amino acids. That sequence of amino acids forms one **polypeptide**. The length of DNA that codes for one polypeptide is called a **gene**.

One gene codes for one polypeptide. This may form a protein. If the protein contains more than one polypeptide (i.e. it has a quaternary structure), there will be more than one gene used to code for the protein.

Protein synthesis

Revised

The code that is carried in the sequence of the bases of a gene is transferred to the structure of a messenger RNA (mRNA) molecule. This is called transcription.

The mRNA passes out of the nucleus via a nuclear pore and attaches to a ribosome that consists of ribosomal RNA (rRNA). The ribosome is the site of **protein synthesis**. Transfer RNA (tRNA) molecules bring up amino acids and align themselves according to the sequence of base triplets on the mRNA. This is known as translation.

The sequence of base triplets therefore dictates the sequence of amino acids that are then joined together to form a polypeptide.

Typical mistake

Some candidates try to learn too much detail about protein synthesis. Don't worry about the details — this comes into A2 biology.

Revision activity

Draw a flow diagram to show the sequence of events in protein synthesis.

Examiner's tip

Ensure that you understand how the structure of DNA and the different forms of RNA enable them to perform their functions.

Exam practice

1 **(a)** Describe the structure of a DNA nucleotide. [3]

(b) State two differences between the nucleotides of DNA and those of RNA. [2]

(c) Explain why RNA molecules are much smaller than DNA molecules. [2]

2 In a well-known experiment, scientists allowed bacteria to grow on a medium that contained only heavy nitrogen. The bacteria incorporated this nitrogen into their DNA. After many generations, the scientists assumed that all the nitrogen in the DNA had been replaced by heavy nitrogen. They then placed the bacteria on a medium containing normal nitrogen and allowed them to breed. The DNA was extracted from the bacteria after successive generations and separated according to its mass. This separation technique involved centrifuging the DNA in a dense solution, which caused bands of DNA to appear in the centrifuge tube. Each band represents DNA of a different mass. Heavier DNA appears in a band lower down the tube. The results are shown in the diagram below.

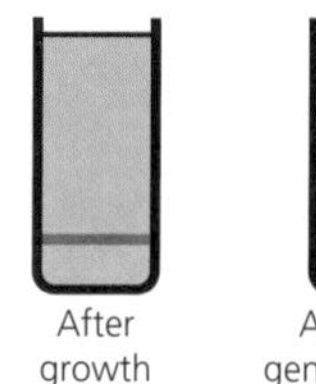
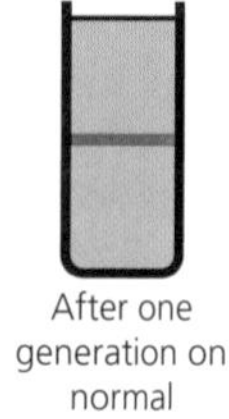
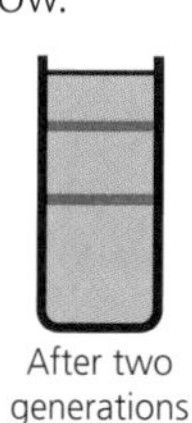
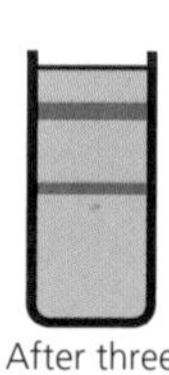

(a) Which part of the DNA contains nitrogen? [1]

(b) The scientists believed that this experiment demonstrated that DNA is replicated in a semi-conservative fashion. Explain what is meant by the term *semi-conservative replication*. [3]

(c) Explain how the results in the diagram demonstrate semi-conservative replication. [4]

(d) If replication had been conservative — where whole molecules of DNA are conserved and totally new ones are made — the results would show a different banding pattern. On the diagram (right), show what banding you would expect to see using DNA collected from bacteria after two generations on normal nitrogen. [2]

After two generations on normal nitrogen

3 The following table shows information about base ratios in three species.

Species	Percentage composition of bases			
	A	T	C	G
X	28.2	29.2	21.0	20.8
Y	24.3		24.9	
Z	34.2		15.8	

(a) State the full names of the bases A, T, C and G. [4]

(b) Complete the table. [2]

(c) Suggest why the total for species X is not exactly 100%. [2]

Answers and quick quiz 8 online

Online

Examiner's summary

By the end of this chapter you should be able to:

- State that the structure of DNA is a double-stranded polynucleotide with four bases: adenine, thymine, cytosine and guanine.
- Describe how the bases pair in a complementary fashion.
- Describe the structure of RNA and how it differs from that of DNA.
- Outline the semi-conservative method of DNA replication.
- Outline the way that DNA codes for the structure of polypeptides.
- Outline the roles of DNA and RNA in protein synthesis.

9 Enzymes

Enzyme activity

Examiner's tip

This is an important topic for a number of reasons:

- it may be tested in the F212 written examination
- it is also likely to be tested as part of the practical unit F213
- it is the underlying principle for many topics that appear in A2 biology and could be tested again as the synoptic element of A2

Enzymes as proteins

Revised

Enzymes are **globular proteins**, so many of their properties are similar to those of protein molecules. They have a specific three-dimensional shape, known as their **tertiary structure**.

A **globular protein** is a protein that folds and coils into a globular shape rather than a fibre.

Enzymes act as **catalysts** to **metabolic reactions** in **living organisms**, which means they usually speed up metabolic reactions so that they occur at a reasonably fast pace even at body temperature.

Typical mistake

Candidates tend to describe enzymes 'breaking down' the substrate. It is far more precise to say 'hydrolyse' or 'oxidise', i.e: to name the actual reaction that occurs.

Enzyme action may be **intracellular** (working inside cells), for example lysins in lysosomes which hydrolyse large organic molecules. Alternatively, enzymes may be **extracellular** (working outside cells), for example digestive enzymes which are released into the digestive system.

How do enzymes work?

Revised

Enzymes have particular properties. These include:

- the molecule has a particular shape
- part of the molecule is an **active site** that is complementary to the shape of the substrate molecule
- each enzyme is specific to the substrate
- there is a high turnover number
- they have the ability to reduce the energy required for a reaction to occur
- their activity is affected by temperature, pH, concentration of enzyme and concentration of substrate
- the enzyme is left unchanged at the end of the reaction

Examiner's tip

Recalling a list of the properties of enzymes is easy. You need to be able to *explain* how those properties are related to the structure of the enzyme molecules.

Specificity and the lock and key hypothesis

The **specificity** of an enzyme refers to its ability to catalyse just one reaction or type of reaction. Only one particular substrate molecule will fit into the active site of the enzyme molecule. This is because of the shape of the active site.

The shape of the active site is caused by the specific sequence of amino acids. This produces a specific tertiary structure — the three-dimensional shape of the molecule. This is referred to as the **lock and key hypothesis**.

The **lock and key hypothesis** explains how enzymes are specific to their substrate.

Catalysing the reaction

Enzymes can speed up the rate of a reaction at body temperature. This is achieved by enabling the reaction to take a slightly different route by forming an **enzyme–substrate complex**. This complex alters the bonding in the substrate, enabling bonds to be broken more easily.

Enzymes lower the **activation energy** required for the reaction to occur. The activation energy is the amount of energy required to set off the reaction and break the bonds in the substrate molecule.

Typical mistake

Candidates tend to refer to the lock and key hypothesis incorrectly. They make statements such as 'the enzyme works by the lock and key method', which is incorrect. The lock and key hypothesis explains how enzymes are specific to their substrate; it does not explain how they work.

Induced-fit hypothesis

The **induced-fit hypothesis** (Figure 9.1) helps to explain how the activation energy may be reduced.

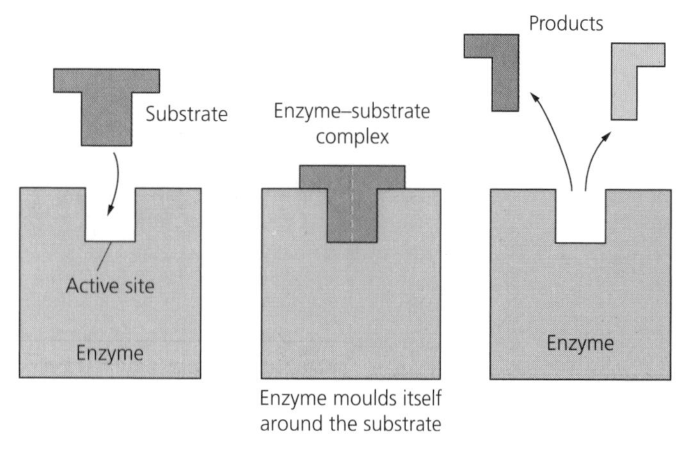

Figure 9.1 The induced-fit mechanism of enzyme action

Now test yourself

1 Using the analogy of a boulder in a hollow at the top of a hill, explain the role of an enzyme helping to overcome the activation energy in a reaction.

Answer on p. 110

Tested

The **induced-fit hypothesis** is a hypothesis that modifies the lock and key hypothesis.

The active site of an enzyme molecule does not have a perfectly complementary fit to the shape of the substrate. When the substrate moves into the active site, it interacts with the active site and interferes with the bonds that hold the shape of the active site. As a result, the shape of the active site is altered to give a perfect fit to the shape of the substrate. This changes the shape of the active site, which also affects the bonds in the substrate, making them easier to make or break.

Typical mistake

Some candidates think that the lock and key hypothesis and the induced-fit hypothesis are mutually exclusive. However, the induced-fit hypothesis is a way to explain how a substrate moves into an active site that then changes to fit the substrate like a lock and key.

Now test yourself

2 Describe what bonds and interactions could cause the enzyme to change shape to wrap around the substrate more closely.

Answer on p. 110

Tested

The course of an enzyme-controlled reaction

In an **enzyme-controlled reaction**, the enzyme and substrate molecules combine to form an enzyme–substrate complex. The substrate is converted to the product, forming an **enzyme–product complex**.

The product is finally released and the enzyme is then free to take up another substrate molecule. This process is shown in Figure 9.2.

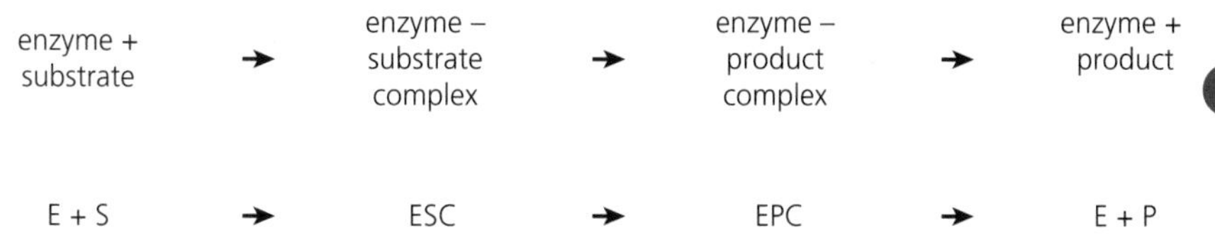

Figure 9.2 An enzyme-controlled reaction

Typical mistake

Many candidates forget to mention the enzyme–substrate complex.

Factors affecting enzyme activity

pH

Revised

All enzymes have an **optimum pH**, so they will not work as quickly at a pH outside their optimum range (Figure 9.3). This is because the hydrogen ions that cause acidity affect the interactions between the *R* groups.

Altering the interactions between the *R* groups affects the tertiary structure of the molecule and may alter the shape of the active site. The shape will no longer be complementary to the shape of the substrate molecule.

The **optimum pH** is the pH at which enzymes work best.

Revision activity

Draw a graph to show the effect that pH change has on the activity of an enzyme. Annotate the graph with explanations for what happens at each stage.

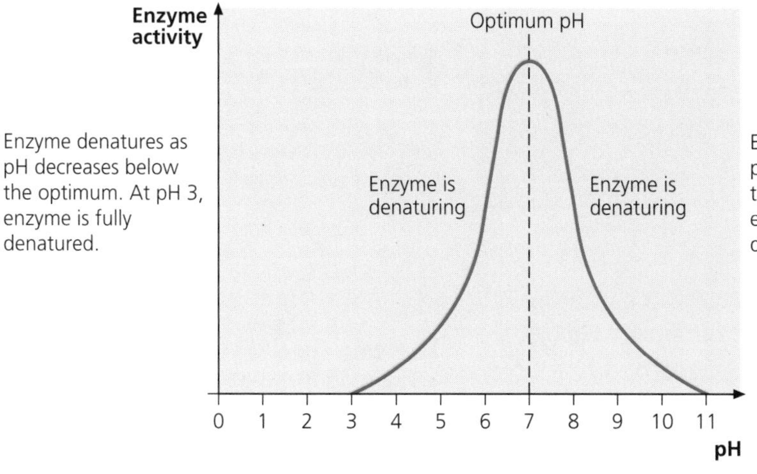

Figure 9.3 The effect of pH on an enzyme-catalysed reaction

Temperature

Revised

The effect of **temperature change** on enzyme action varies depending on the temperature range considered (Figure 9.4). Each enzyme has an optimum temperature at which it is most active. This temperature is often 37°C (in mammals), but it may be different in other organisms.

At low temperatures (0–45°C) the activity of most enzymes increases as temperature rises. This is because of the effect of temperature on the kinetic energy of the molecules. At low temperatures, the molecules have little kinetic energy. They collide infrequently with the substrate molecules

Typical mistake

Many candidates seem to think that all enzymes have the same optimum temperature and pH, which is not the case.

and activity is reduced. As temperature rises, the molecules gain more kinetic energy. They collide more frequently with the substrate molecules and so are more likely to have sufficient energy to overcome the required activation energy. Therefore, activity increases.

At higher temperatures, enzymes will lose their shape (they become denatured). Higher temperatures cause increased vibration of parts of the molecule. If the temperature rises above a certain point, the bonds within the enzyme molecule vibrate too much and break, which alters the bonding in the active site, changing its shape. The active site no longer fits the shape of the substrate and activity reduces quickly to zero.

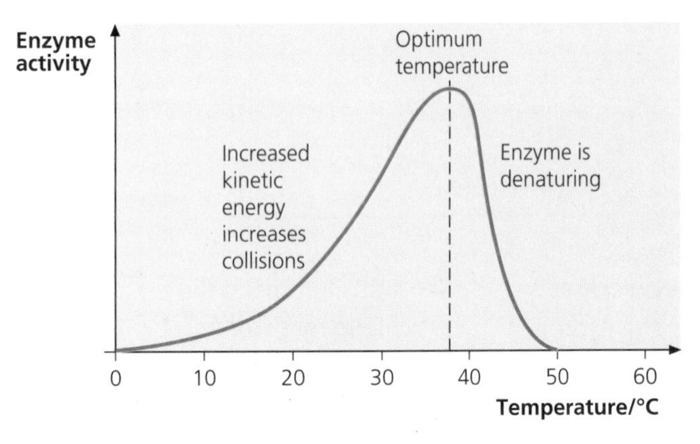

Figure 9.4 The effect of temperature on the rate of an enzyme-catalysed reaction

Typical mistake

When describing or explaining the effects of changing conditions on enzyme action, many candidates make statements such as 'temperature affects enzyme activity'. They forget to say that *changing* the temperature affects enzyme activity.

Revision activity

Draw a graph to show the affect that temperature change has on the activity of an enzyme. Annotate the graph with explanations for what happens at each stage.

Enzyme concentration

Revised

If there are more enzyme molecules in a particular volume of reaction medium, there are more active sites available. There is a greater likelihood of collisions between the enzyme and the substrate molecules. More interactions per second means there is a higher rate of reaction. Therefore, as the **enzyme concentration** increases, so too does the rate of reaction (Figure 9.5).

The effect of temperature change and pH on reaction rates is usually because at extremes of temperature and pH some enzyme molecules are denatured and the concentration of active enzyme molecules is reduced.

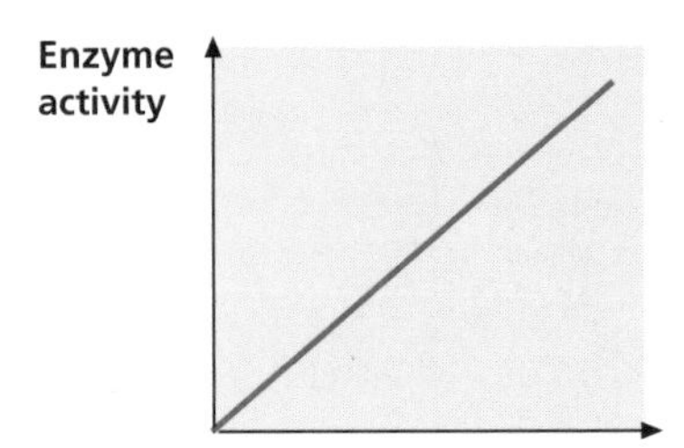

Figure 9.5 The effect of enzyme concentration on enzyme activity

Substrate concentration

Revised

If the **substrate concentration** is high, there is a greater chance of collisions between the enzyme and the substrate molecules. Therefore, as the substrate concentration increases, so too does the rate of reaction (Figure 9.6).

If the number of enzyme molecules is limited, the rate of reaction levels off once all the enzyme active sites are fully occupied.

Examiner's tip

Questions could be limited to just one property, in which case it is likely that a lot of detail will be expected. Alternatively, this topic could be tested with a more open-ended type of question, in which case less detail will be expected on each property.

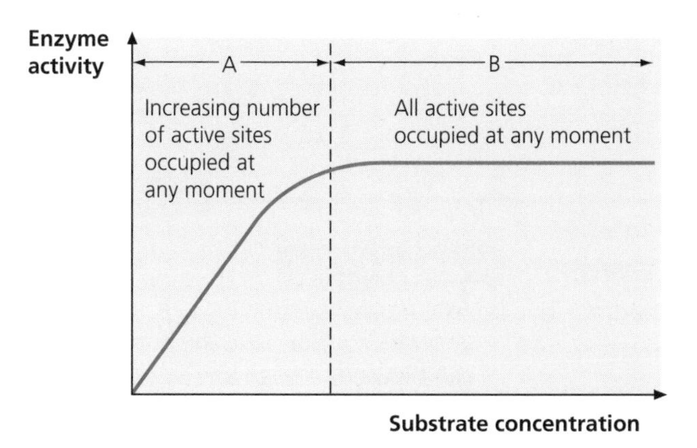

Figure 9.6 The effect of substrate concentration on enzyme activity

Revision activity

Draw a mind map to link together all the factors that affect enzyme activity and the explanations for each.

Cofactors and coenzymes

Revised

Cofactors are inorganic substances, usually metal ions. They fit into the active site and activate the enzyme. For example, the enzyme catalase contains iron ions.

Coenzymes are larger organic substances that take part in the reaction. They usually transfer other reactants between enzymes. For example, coenzyme A takes part in aerobic respiration.

Cofactors are substances (usually a small metal ion) that are required to make enzymes function correctly.

Coenzymes are large organic substances that are involved in some chains of enzyme-controlled reactions.

Inhibitors

Revised

Inhibitors are substances that reduce the rate of reaction and fit into a site on the enzyme.

Inhibitors are substances that reduce the activity of an enzyme.

Competitive inhibitors

Competitive inhibitors have a shape similar to the shape of the substrate. They fit into the active site, stopping the substrate molecules fitting in. This reduces the number of available active sites. The amount of inhibition depends on the relative concentrations of inhibitor and substrate molecules. Some **medicinal drugs** work by inhibiting the activity of enzymes.

Now test yourself

3 Explain how a non-competitive inhibitor prevents the substrate entering the active site.

Answer on p. 110

Tested

Non-competitive inhibitors

Non-competitive inhibitors fit into a different site on the enzyme molecule. They cause a change in the shape of the enzyme molecule. This affects the active site, so the substrate molecule can no longer fit in.

Reversible and irreversible inhibitors

Non-reversible inhibitors bind permanently to the enzyme. **Reversible inhibitors** occupy the enzyme site only briefly.

Typical mistake

Many candidates think that competitive inhibitors are reversible and non-competitive inhibitors are irreversible, but this is not the case.

Drugs and poisons

Many **poisons** act by inhibiting enzymes. A poison such as cyanide inhibits the action of the enzyme cytochrome oxidase in aerobic respiration. Cytochrome oxidase contains iron ions and the cyanide binds to them. Many medicinal drugs also act as inhibitors of enzymes in the body. This is why taking the correct dose of medicinal drugs is important as overdosing can be lethal, especially if the inhibitor is non-reversible.

Practical skills

How Science Works

Revised

The written examination will include a number of marks for How Science Works. These are often set in the context of a practical test carried out by a student or scientist. The action of enzymes is likely to be involved. You should be familiar with how the effects of changes in the following factors on enzyme activity can be investigated experimentally:

- pH
- temperature
- enzyme concentration
- substrate concentration

It is important to consider the following points.

1. *Volume and concentration of enzyme solution.*
2. *Volume and concentration of substrate solution.*
3. *Control of temperature.* A thermostatic water bath is often the best way.
4. *Control of pH.* A buffer solution controls pH.
5. *How can you make the test more reliable?* Repeat the test a number of times and calculate the mean.
6. *Testing reliability.* Comparing raw data to the mean is a good indication. There should be little variation around the mean. Calculating the standard deviation is even better.
7. *What is the control?* This is a test that omits one factor in the experiment to show that it is essential for the reaction to occur. It is usual to omit the enzyme from the reaction mixture to show that no reaction occurs without the enzyme.
8. *What level of precision is appropriate in the measurements?*
9. *How valid is the experiment?* Is it actually measuring what you think it should? Have you taken account of all possible conditions that may affect the reaction rate?

Examiner's tip

As enzyme experiments can generate a lot of data, this topic can be used to test How Science Works, particularly the aspects involving interpretation of data and evaluation of both methods and results.

Examiner's tip

To gain full marks you must refer to the concentration and the volume of solutions.

Typical mistake

Many candidates refer to the amount of a substance when describing investigations. Ask yourself, what is the unit of 'amount'?

Revision activity

Sketch the apparatus needed to collect and measure the volume of gas released from decomposing hydrogen peroxide. Annotate the diagram with information about how each variable can be controlled.

Now test yourself

Tested

4. Explain why it is important that only one variable is changed in a practical test.
5. Explain the difference between controlling a variable and the use of a control in a practical test.

Answers on p. 110

Exam practice

1 Amylase is an enzyme that converts starch (amylose) to reducing sugars. A student investigated the effect of different temperatures on the rate of action of amylase. She prepared water baths containing starch suspension at four different temperatures. She then added amylase to each sample of starch and stirred. Each minute, the student transferred two drops of the starch/amylase mixture to a cavity tile containing iodine solution and noted the colour produced. The results are shown in the following table.

Time (min)	Temperature (°C)			
	5	20	40	70
1	Blue/black	Blue/black	Yellow	Blue/black
2	Blue/black	Blue/black	Yellow	Blue/black
3	Blue/black	Blue/black	Yellow	Blue/black
4	Blue/black	Dark yellow	Yellow	Blue/black
5	Blue/black	Yellow	Yellow	Blue/black
6	Blue/black	Yellow	Yellow	Blue/black
7	Blue/black	Yellow	Yellow	Blue/black

(a) **(i)** What does the colour blue/black indicate? [1]

(ii) What does the yellow colour in the samples at 20°C and 40°C indicate? [1]

(b) Suggest why the tube at 40°C was yellow after 1 minute. [2]

(c) Explain why the sample at 5°C remained blue/black. [3]

(d) Explain why the sample at 70°C remained blue/black. [4]

(e) As part of her evaluation, the student commented that it was difficult to sample all the tubes at the same time. Suggest one improvement she could make to her experiment. [1]

(f) She concluded that the optimum temperature for amylase activity is 40°C. Explain why this figure may not be accurate and suggest further improvements to her procedure that may make the test more accurate. [3]

2 **(a)** Explain what is meant by the term *biological catalyst*. [2]

(b) Explain why more than one enzyme is needed to digest starch. In your response, ensure that the properties of enzymes are linked to their structure. [7]

3 **(a)** A student performed a practical procedure in which the rate of formation of maltose was measured in the presence and absence of chloride ions. In the presence of chloride ions, the rate of maltose formation increased.

(i) State the name given to a metal ion that increases the rate of an enzyme-controlled reaction. [1]

(ii) Suggest how the chloride ions act to have this effect on the rate of reaction. [2]

(b) The student extended his investigation to test the effect of temperature on the rate of reaction. When explaining his results, he made the following statement:

As the heat increased, the reaction went faster until it got to its highest. After this, the rate of reaction fell. This happened because the enzyme was killed and the hydrogen peroxide could not fit into the enzyme's key site.

Suggest a more appropriate word to replace each of the underlined words. [4]

Answers and quick quiz 9 online

Online

Examiner's summary

By the end of this chapter you should be able to:

- ✔ State that enzymes are globular proteins with a specific tertiary structure, which catalyse metabolic reactions.
- ✔ State that enzyme action may be intracellular or extracellular.
- ✔ Describe the mechanism of action of enzyme molecules, with reference to specificity, active site, lock and key hypothesis, induced-fit hypothesis, enzyme–substrate complex, enzyme–product complex and lowering of activation energy.
- ✔ Describe and explain the effects of changes in pH, temperature, enzyme concentration and substrate concentration on enzyme activity.
- ✔ Explain the effects of competitive and non-competitive inhibitors on the rate of enzyme-controlled reactions, with reference to both reversible and non-reversible inhibitors.
- ✔ Explain the importance of cofactors and coenzymes in enzyme-controlled reactions.
- ✔ State that metabolic poisons and some medicinal drugs may be enzyme inhibitors.

10 Diet and food production

A balanced diet

Components of a balanced diet

Revised

A **balanced diet** supplies all the requirements for health. There are seven components in a balanced diet:

- macronutrients — required in large amounts:
  - proteins
  - carbohydrates
  - fats
- micronutrients — required in small amounts:
  - minerals
  - vitamins
- other essential components:
  - fibre
  - water

These components should not be consumed in excess otherwise the diet will become unbalanced, leading to **malnutrition**.

The total amount of energy in a balanced diet is the key to avoiding **obesity**. This means that energy **consumption** must be balanced with energy use. As soon as energy consumption is higher than energy use, the person's weight will increase which may lead to obesity. If energy consumption is lower than energy use, weight will be lost. Energy is gained from fats, carbohydrates and proteins in the diet.

A **balanced diet** is one that supplies all the necessary components for a healthy life in the correct proportions.

Examiner's tip

Obesity is big news in Britain and in much of the Western world, so there is plenty of data out there to research. This topic could be used to test How Science Works, notably interpreting data. Make sure you practise dealing with data.

Diet and coronary heart disease (CHD)

Revised

Coronary heart disease (CHD) is a disease of the coronary arteries. Many factors contribute to the development of CHD, including age, high blood pressure (hypertension), high blood cholesterol level, gender, stress, diet, body mass, hereditary factors, exercise and smoking. No single factor can be said to 'cause' CHD — it is a multifactoral disease.

However, one of the main factors involved in CHD is blood cholesterol level, which is affected by diet. Healthy people are advised to keep their blood cholesterol level to below 5.0 $m/mol/dm^{-3}$.

Coronary heart disease (CHD) is a disease of the coronary arteries that reduces the flow of blood to the heart muscles.

Revision activity

Draw a mind map with CHD in the centre. Note the factors that contribute to CHD and include notes explaining how each factor contributes to this disease.

Typical mistake

Candidates often suggest that one factor such as high blood cholesterol causes CHD.

Now test yourself

Tested

1 Explain why CHD is considered to be a multifactoral disease.

Answer on p. 110

What is blood cholesterol?

Fats and cholesterol are not soluble in water. They are transported as **lipoproteins** in the blood.

Lipoproteins may be **high density lipoproteins (HDLs)**, which are sometimes called 'good' cholesterol, or they may be **low density lipoproteins (LDLs)**, which are sometimes called 'bad' cholesterol. LDLs in the blood lead to greater deposition in the artery walls because the artery walls contain LDL receptors that get exposed when the wall is damaged. The ratio of HDL to LDL is important. The LDLs should make up no more than two-thirds of the total lipoproteins in the blood.

Examiner's tip

Be specific with your descriptions. For example, always use the term *blood cholesterol level* rather than *cholesterol level*.

Links between fat consumption and CHD

A high fat diet can raise blood cholesterol levels. Saturated fats (animal fats) are carried in the blood as low density lipoproteins, so a diet high in saturated fats increases the concentration of LDLs in the blood. High concentrations of LDLs in the **circulatory system** lead to greater deposition of fatty substances in the artery walls. This is called **atherosclerosis**.

Atherosclerosis is the deposition of fatty substances in the walls of the arteries.

Polyunsaturated fats or oils come from plants and oily fish. These fats are carried in the blood as high density lipoproteins. A high proportion of unsaturated fats in the diet can help to reduce deposition in the artery walls.

Examiner's tip

There are lots of details about how the relative levels of HDL and LDL affect the health of the circulatory system — so be ready.

Atherosclerosis

High blood pressure can damage the endothelium (inner lining) of the artery. This exposes the LDL receptors in the wall of the artery. LDLs in the blood attach to the receptors and release their cholesterol. This deposition occurs in the wall of the artery, under the endothelium.

Examiner's tip

Remember how to spell *atherosclerosis* correctly and don't confuse it with *arteriosclerosis*.

Cholesterol and other fatty substances build up and thicken the wall of the artery, which narrows the lumen. The narrowed artery carries less blood and therefore less oxygen to the muscles of the heart. Oxygen is needed for aerobic respiration and a lack of oxygen causes angina.

The deposition and narrowing of the arteries also increases blood pressure further.

Examiner's tip

The sequence of events leading to deposition of materials in the artery walls could be used to test quality of written communication.

Refer to deposition 'in the wall of the artery' rather than 'deposition in the artery' or 'on the wall of the artery'.

Typical mistake

Many candidates state that CHD reduces the flow of blood to the heart. This could refer to the blood returning in the veins to the atria, so it is important to refer to reduced blood flow to the heart muscle.

Now test yourself

Tested

2 Explain the role of high blood pressure in contributing to atherosclerosis.

Answer on p. 110

Revision activity

Draw a flow chart describing how atherosclerosis occurs.

Other dietary effects

In addition to fat consumption, diet has other effects that link it to CHD:

- excess salt in the diet can raise blood pressure
- excess carbohydrates and sugars may lead to obesity or diabetes, which increase the risk of CHD
- high levels of cholesterol in the diet are likely to raise blood cholesterol levels

Availability of food

Almost all energy used in food chains originates from the sun. Plants convert the sun's energy into chemical energy in organic molecules. These molecules, such as sugars and fats, are the basis of our food.

Examiner's tip

You are not expected to know any details about **food chains**, just the principle that all our food comes from plants or animals that eat plants.

Production of food

Revised

Scientific ideas and techniques employed to increase food production include:

- selective breeding
- fertilisers
- pesticides
- antibiotics
- microorganisms

Examiner's tip

You are not expected to know a great amount of detail about each of these techniques. It is, however, important that you understand the scientific principles behind each method.

Selective breeding

Plants

Selective breeding is used to produce **crop plants** with a number of desired qualities, including:

- disease resistance
- pest resistance
- high yields
- ease of harvest

Selective breeding is the process of breeding plants and animals for particular traits.

The process is simple and has been used for thousands of years:

1. Select examples of the plants that display the desired characteristic, for example plants that have a high yield.
2. Breed two such high yield plants together and allow them to produce seeds.
3. Plant the seeds so that they germinate and grow into new plants. There will probably be a wide variety of yields among the offspring.
4. Select those offspring that show greater yields.

5 Breed them together again.
6 Continue this process for many generations.

Breeders are also able to produce a strain that has a combination of desired characteristics. This can be achieved by selecting plants from other closely related strains of the same species:

1 Select examples of the plants that display the desired characteristic, for example plants that have a high yield and others that are resistant to fungal attack.
2 Breed two such strains together and allow them to produce seeds.
3 Plant the seeds so that they germinate and grow into new plants.
4 Test the new plants for their ability to produce high yields and resist fungal attack.
5 Select those plants that display the required combination.
6 Breed them together again, or breed with a high yield strain again.
7 Continue this process for many generations.

Animals

Selective breeding in **domestic animals** involves the same process as that for plants. The desired characteristics include:

- high yields of meat, milk or eggs
- high quality meat
- high or low fat milk

Fertilisers

As crops are grown and harvested, they remove minerals from the soil. These minerals must be replaced, otherwise the soil loses fertility. Certain minerals are particularly important:

- nitrates are required for the plants to make protein
- phosphates are required for making DNA and for metabolic reactions in the cells
- potassium is required for protein production and healthy growth
- calcium is important in making cell walls

Modern farming includes soil testing to monitor the levels of mineral ions in the soil. **Fertilisers** are applied in appropriate amounts to reduce wastage and to avoid polluting surrounding watercourses.

Pesticides

In general, pests will reduce crop yields in some way, such as by:

- competing for resources
- reducing plant growth due to disease
- damaging the crop, making it unmarketable or less valuable

Fungicides, insecticides, bacteriocides and herbicides are all types of **pesticide**. Each pesticide has a specific role in reducing the effect that a particular pest can have on the crop and, therefore, increasing yields.

Antibiotics

Disease in animals will reduce yield. This is because the:

- animals expend energy fighting the disease, so less energy goes into growth

Typical mistake

Many candidates say that selection causes variation or mutation in the offspring. However, variation occurs due to mutations that are random. This variation allows the breeder to select the best varieties.

Typical mistake

Many candidates fail to state that the selection process must occur over a large number of generations.

Now test yourself

3 Describe how a breeder can produce cattle that give a high milk yield low in saturated fat.

Answer on p. 110

Tested

Typical mistake

Many candidates fail to link the underlying biology, such as selection, to how we increase food production.

- disease may reduce the final weight of the animal, and therefore the yield
- disease may lower the production of milk, eggs, etc.

Some diseases can be passed on to humans and so may necessitate the destruction of the animal at a loss to the farmer.

Bacterial and fungal diseases can usually be treated with **antibiotics**, which help the animal to recover more quickly and reduce the chance of the disease spreading to others.

Microorganisms

The main reasons for using **microorganisms** to make food are related to how quickly they can grow, reproduce and manufacture products. In the right conditions, a colony of microorganisms can double in size every 30 minutes. This growth can be optimised by using a fermenter, which is a specialised vessel that monitors the conditions required for growth and keeps all the conditions precisely as needed. This includes:

- maintaining the required temperature
- keeping the pH constant
- supplying nutrients and oxygen
- removing products and wastes

As a result, food production can be rapid — protein for **human consumption** can be grown much more quickly than by rearing animals or growing plants on a farm or in a factory setting. The type of food grown is called single-cell protein.

Table 10.1 Advantages and disadvantages of using microorganisms

Advantages	Disadvantages
Production can be increased and decreased fairly quickly to match demand No animals are used, so vegetarians can eat the proteins There are no issues over how animals are reared The protein produced is healthier as it contains: • all the essential amino acids (it is a first-class protein) • less fat • fewer saturated fats • no cholesterol The microorganisms can digest and grow on almost anything organic, so they can be grown on waste products from other processes or even on domestic waste The protein produced can be much cheaper than animal-derived protein	Many people may not want to eat single-cell proteins that have been grown by fungi. They may feel it is unclean or contaminated People may not want to eat protein that has been grown on waste products The protein does not have the same taste or texture as traditional animal protein The protein contains less energy The protein needs to be isolated and purified before use The fermenters used maintain ideal conditions for microorganism growth. They may become infected with the wrong microorganisms or with pathogenic microorganisms. Such unwanted microorganisms will also grow quickly, making the production process less efficient or the product unusable

Examiner's tip

You are not expected to know a lot of detail, but expect many questions to test the application of your understanding in unfamiliar contexts. Don't be thrown by the context — think carefully about what scientific principles can be applied.

Now test yourself

4 Farmers used to put antibiotics in animal feed, regardless of whether or not the animal needed it. Suggest why this practice developed and why it has now been banned.

Answer on p. 110

Tested

Revision activity

Draw a fermenter and annotate the diagram with notes explaining how and why the conditions are maintained.

Preventing food spoilage

Revised

How do microorganisms spoil food?

Food is spoilt by the action of microorganisms, which are everywhere so it is difficult to stop them getting into our food. As microorganisms grow, they release enzymes that digest the food. They may also release harmful waste products that act as toxins.

Preservation techniques

Food preservation relies on keeping the food in conditions that do not allow the microorganisms to breed and grow. Techniques to prevent **food spoilage** (Table 10.2) can be classified according to how they work. These work by:

- killing the microorganisms
- preventing further microorganisms getting on the food
- inhibiting the growth and reproduction of any microorganisms that are already on the food

Typical mistake

Many candidates fail to make the link to show *how* the microorganisms are killed or prevented from reproducing.

Table 10.2 Methods of preventing food spoilage by microorganisms

Method	Technique	Scientific principle
Killing the microorganisms	Cooking	The heat denatures the enzymes and other proteins in the microorganisms, which kills them
	Pasteurising	The heat denatures the enzymes in the harmful microorganisms
	Irradiation	This damages the proteins and DNA in the microorganisms, killing the microorganisms or preventing their reproduction
Preventing further microorganisms getting on the food	Canning or bottling Vacuum wrapping Any plastic or paper packaging	These forms of packaging prevent microorganisms getting on the food after it has been treated to kill those already present
Inhibiting the growth and reproduction of any microorganisms that are already on the food	Cooling or freezing	This does not kill the microorganisms present, but at low temperatures their enzymes work slowly. This slows growth and reproduction
	Drying, salting or adding sugar	These processes use osmosis to dehydrate any microorganisms present and prevent their action on the food
	Pickling	This provides an acidic environment that denatures the enzymes and prevents microorganisms growing
	Smoking	This leaves the food with a hardened and dry outer surface that acts osmotically. Smoke also contains antimicrobial chemicals that can reduce the action of the microorganisms or even kill them

Revision activity

Find a number of pictures showing food in its packaging. Annotate each one with the type of treatment used for preservation and explain how each technique works.

Examiner's tip

You are not expected to know a great amount of detail about each of these techniques. It is, however, important that you understand the scientific principles behind each method and how it keeps the food safe to eat.

Now test yourself

Tested

5 Explain why it is essential to package food after it has been treated to kill microorganisms.

Answer on p. 110

Exam practice

1 The graph below shows the percentage of children who were obese in England and Wales between 2000 and 2010.

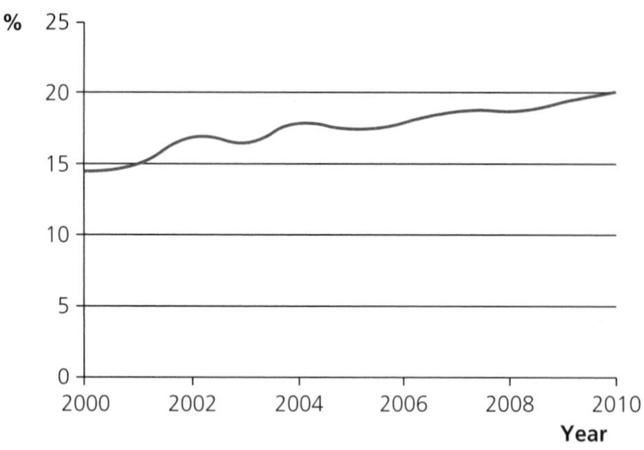

(a) Describe the trend shown by the graph. [2]

(b) State three components of diet that may contribute to obesity. [3]

(c) Explain how obesity can contribute to atherosclerosis. In your answer you should make clear the sequence of events in this process. [7]

2 **(a)** List three conditions required for microorganisms to grow rapidly. [3]

(b) **(i)** Chlorella is a microscopic green alga. It can be grown on almost any organic matter and it divides every 30 minutes in ideal conditions. Over 60% of its dry mass is proteins and it contains all the essential amino acids. Describe and explain the advantages of using chlorella to manufacture single-cell protein. [7]

(ii) Explain why it is important to ensure that the tanks used to grow chlorella do not become infected with other microorganisms. [3]

Answers and quick quiz 10 online

Online

Examiner's summary

By the end of this chapter you should be able to:

- ✔ Define the term *balanced diet*.
- ✔ Explain how consuming an unbalanced diet can lead to malnutrition.
- ✔ Discuss possible links between diet and coronary heart disease (CHD).
- ✔ Discuss the effects of high blood cholesterol levels with reference to low density lipoproteins (LDLs) and high density lipoproteins (HDLs).
- ✔ Explain how humans depend on plants for food.
- ✔ Describe how selective breeding, fertilisers, pesticides and antibiotics can be used to increase food production.
- ✔ Describe the advantages and disadvantages of using microorganisms to produce food.
- ✔ Outline how food can be preserved.

11 Health and disease

Health and disease

Health is complete physical, mental and social wellbeing. **Disease** is any condition in which you do not have complete physical, mental and social wellbeing. It may be caused by infection, a malfunction of the body or mind, poor diet, stress or a poor social situation. Many diseases are caused by **parasites** or **pathogens**.

A **parasite** is an organism that lives on or in another organism (its host) and gains nutrition from the host.

A **pathogen** is a microorganism that can cause disease.

Examiner's tip

You should expect to see questions testing aspects of How Science Works. In particular, you may be tested on science in a social context or use of data about the spread of a disease.

Malaria

Revised

Malaria is caused by the eukaryotic organism *Plasmodium*. There are several species of *Plasmodium*, including *P. vivax*, *P. falciparum* and *P. malariae*.

Plasmodium lives and breeds inside the red blood cells and liver cells of the human. It feeds on the haemoglobin in the red blood cells. It is spread by a vector — an organism that transmits the microorganism from an infected person to an uninfected person. The vector is the female *Anopheles* mosquito.

Hundreds of millions of people are infected with the malaria parasite and many are newly infected each year. Estimates of the numbers of worldwide deaths per year vary between 0.66 and 1.2 million people.

It is difficult to make vaccinations for malaria as the pathogen lives much of its life inside body cells. It also alters the **antigens** on its surface frequently.

Typical mistake

Many candidates are not careful enough about the details — the pathogen that causes malaria is transmitted by the female *Anopheles* mosquito, not just by a mosquito.

Antigens are molecules on the surface of cells that trigger an immune response.

Examiner's tip

Remember that it is the pathogen that is transmitted, not the disease so people are infected by *Plasmodium* not malaria.

Now test yourself

Tested

1 Using the transmission of malaria as an example, explain the difference between pathogens, parasites and diseases.

Answer on p. 111

HIV/AIDS

Revised

Human immunodeficiency virus (**HIV**) is the virus responsible for acquired immune deficiency syndrome (**AIDS**). When a person is infected by HIV, proteins in the membrane surrounding the virus bind to specific proteins on the surface of a type of white blood cell known as a T helper cell.

Viruses reproduce by using the host cell to manufacture more viruses, which erupt from the dead host cell and infect more cells. As more T helper cells are killed, the immune system is weakened and the person loses the ability to combat other infections. This is AIDS.

HIV is transmitted from an infected person to an uninfected person by:

- blood to blood contact
- sharing or reusing infected hypodermic needles
- using infected and unsterilised surgical instruments
- accidental needle stick from a discarded syringe
- semen or vaginal fluid during unprotected sexual intercourse
- mother to baby during birth
- breast milk or blood from cracked nipples during breast feeding

Examiner's tip

Make sure that you are specific about the means of transmission — it must be from unprotected sexual intercourse or from infected shared needles.

About 1.8 million people die from AIDS-related diseases every year and approximately 2.7 million people become newly infected. The largest numbers of infected people are in sub-Saharan Africa, but in recent years Eastern Europe has become a centre of infection.

There are about 3.4 million children living with HIV/AIDS and some 20 million children have been orphaned by HIV/AIDS. The condition retards economic growth and increases poverty. Antiretroviral drugs, which can delay the onset of AIDS in people who are HIV-positive, are expensive and only widely available to people in more economically developed countries. It is difficult to make vaccinations to prevent HIV as it lives inside body cells.

Tuberculosis

Revised

Tuberculosis (TB) is caused by the bacterium *Mycobacterium*. There are several species including *M. tuberculosis* and *M. bovis*. The disease usually affects the lungs, but it may affect other parts of the body. The pathogen is spread by droplet infection.

TB is spread most easily where people live in crowded conditions, often with poor ventilation. It is associated with poverty and the vast majority of deaths from TB are in the developing world. It is estimated that about one-third of the world's population has been infected with *M. tuberculosis*. However, the majority of these infections do not develop into the disease.

An estimated 14 million people worldwide are infected with active tuberculosis. In 2009, there were 9.4 million new cases of TB and 1.7 million deaths. In the same year, an estimated 380,000 deaths from TB were among people with HIV.

Revision activity

Draw a mind map with one of these diseases (malaria, HIV/AIDS or TB) in the centre and include all the information about its causes, transmission and global impact. Leave space around the diagram to add in any possible links to new medicines, to primary and secondary defences used to combat the disease, and to information about vaccination. Do the same for the other two diseases.

Now test yourself

Tested

2 Explain why TB might be on the increase in inner cities in the UK and why medical professionals are concerned about a link between HIV and TB.

Answer on p. 111

Defence against disease

Defence against disease takes two forms:

- **primary defences**, which prevent the entry of pathogens into the body
- **secondary defences**, which combat pathogens that have already entered the body

Primary defences

Revised

The skin is the main primary defence against pathogens and parasites. It provides a barrier to the entry of microorganisms. The sebum secreted by the skin has antimicrobial properties.

Where the skin is incomplete, other defences reduce the effect of pathogens or parasites that may otherwise gain entry:

- Pathogens entering the digestive system via the mouth will pass into the stomach where a low (acidic) pH will kill almost any living things.
- Pathogens entering the gaseous exchange system are trapped in mucus. This mucus is moved to the top of the trachea and enters the digestive system, where it is passed to the stomach.
- The ears and eyes are protected by the antimicrobial properties of ear wax and tear fluid.
- The urethra and anus are regularly cleaned as materials pass out of the body.
- The vagina is protected by an acidic environment that reduces microbial action and the cervix is blocked by a mucus plug that prevents pathogens entering the female reproductive system.

Secondary defences

Revised

The **immune response** is the body's response to invasion by pathogens.

Non-specific response

The non-specific response involves neutrophils (white blood cells) that engulf and destroy any non-self cells such as dead cells, harmful foreign particles and bacteria. These non-self cells are recognised because they have antigens on their cell surface membranes that trigger the immune system. This is known as phagocytosis and it follows a particular sequence (Figure 11.1):

1. The bacteria are engulfed by the neutrophil.
2. They are surrounded by a vacuole called a phagosome.
3. Lysosomes fuse to the phagosome.
4. Lytic enzymes are released into the phagosome.
5. The bacteria are hydrolysed (digested) and the nutrients can be reabsorbed into the cell.

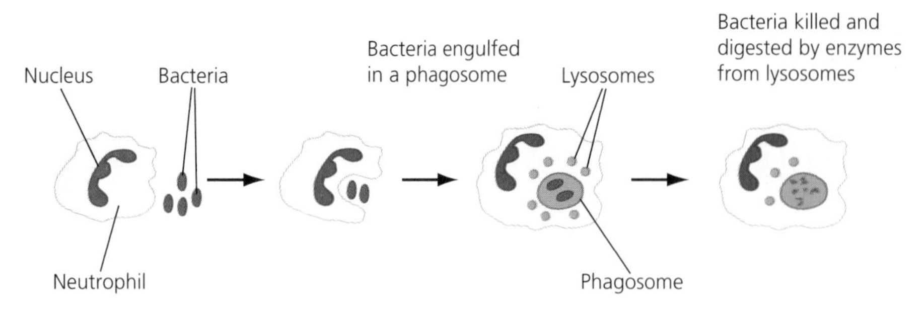

Figure 11.1 The stages involved in phagocytosis

Specific response

The body's immune system can determine which antigens are part of its own body structure. These antigens are known as self-antigens and they do not stimulate an immune response. Non-self antigens are foreign substances that do stimulate an immune response.

The presence of a non-self antigen can trigger the production of **antibodies**. These are special proteins that are secreted by B cells, a type of lymphocyte. These B cells have specific receptors on their cell surface membranes that will only bind to a specific antigen.

Antibodies are proteins that are secreted in response to stimulation by the appropriate antigen. They have specific binding sites and are capable of acting against the pathogen.

T cells are lymphocytes that are also involved in the specific response, but they do not secrete antibodies. Some T cells continually monitor the body for infection and are responsible for killing cells that have been infected with pathogens. Other T cells are involved in the response of the B cells.

The specific response is more complex than the non-specific response and involves a series of stages (Figure 11.2):

1. *Presentation of antigens.* Macrophages are a type of phagocytic cell that have developed from the monocytes in the blood. When these macrophages engulf the bacteria, they do not fully digest them. They isolate the antigens from the surface of the bacteria and present them on their surface membranes. The macrophages become antigen-presenting cells. They travel to the lymph nodes in search of the correct **B cells** and **T cells**. This is a form of **cell signalling** in which the antigen-presenting cells are signalling the presence of a foreign antigen.
2. *Clonal selection.* There may be only a few lymphocytes that carry the correct receptors to bind with the specific antigen. Normally, there will be a B cell, a T helper cell and a T killer cell. These cells must be found and activated. They are selected and activated by coming into contact with the specific foreign antigen. This completes the cell-signalling role of the antigen-presenting cell. The action of the macrophages makes this selection more likely.
3. *Clonal expansion and differentiation.* Once activated, the specific lymphocytes increase in numbers by mitosis. There will normally be a clone of B cells, a clone of T helper cells and a clone of T killer cells. Cells from the clone of B cells will differentiate into plasma cells and **memory cells**. Cells from the T helper cell clone will differentiate into

Typical mistake

Many candidates seem to confuse antigens and antibodies.

Examiner's tip

This is a good area in which quality of written communication can be tested — the examiner can reward candidates who write these steps in the correct sequence.

T helper cells and memory cells. The T helper cells secrete hormone-like substances called cytokines that help to stimulate the B cells. These cytokines are cell-signalling molecules. Cells from the T killer cell clone will become T killer cells. These cells can attack and kill host body cells that are expressing the specific antigen of the pathogen. This expression only occurs when the host cells are infected. Memory cells are long-lived and remain in the blood for some time, providing immunological memory or long-term immunity. If the same pathogen invades again, it will be recognised and attacked more quickly.

4. *Antibody production.* The plasma cells are short-lived. They are specialised to manufacture antibodies. They possess a lot of ribosomes, rough endoplasmic reticulum and Golgi apparatus so that they can manufacture and secrete antibodies quickly.

Typical mistake

Many candidates describe antibodies being produced by lymphocytes. Be specific — antibodies are made by the plasma cells.

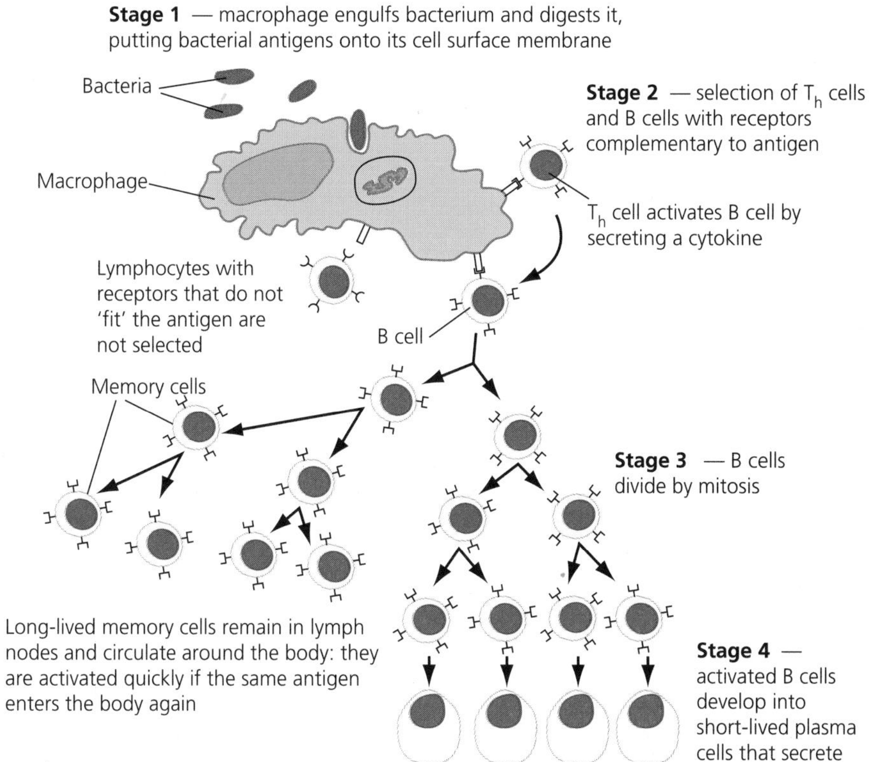

Figure 11.2 The stages of an immune response in which antibodies are produced

Revision activity

Try to visualise this whole specific response process. Draw a simple flow diagram following the process from entry of foreign cells into the body, through their recognition by macrophages and the selection of T and B cells, to the production of specific antibodies that can combat the invading organism.

Now test yourself

3. **Explain why pathogens have antigens that allow them to be recognised as foreign on their surface.**

Answer on p. 111

Tested

Antibodies

Revised

Antibodies are large globular proteins. Most antibodies are similar in structure and possess the same basic components. They consist of four polypeptides held together by disulfide bridges (Figure 11.3).

One end of an antibody is known as the constant region. This has a binding site that can be recognised by phagocytes. It may help binding of phagocytes. The opposite end is known as the variable region. This has binding sites that are specific to a particular antigen. They have a shape that is complementary to the shape of the antigen and can bind to that antigen. The antibody molecule also has a hinge region that allows some flexibility to enhance binding to more than one pathogen.

Examiner's tip

Remember to relate structure to function — this is another possible quality of written communication mark.

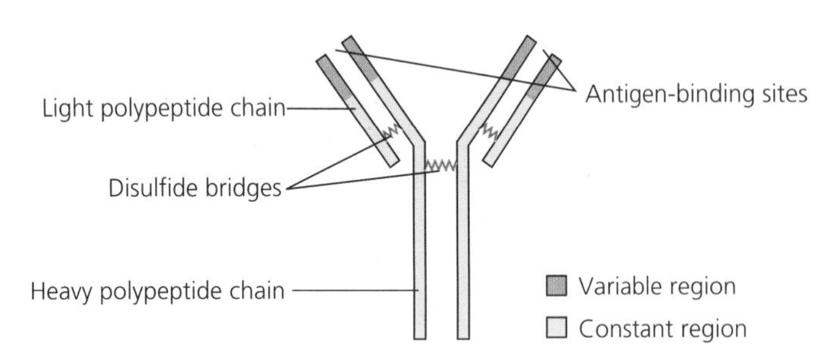

Figure 11.3 The structure of an antibody molecule

The action of antibodies

Antibodies can act on pathogens in a number of ways. These include:

- **agglutination** — the antibody binds to two or more pathogens so that they are held together. This prevents them entering cells and reproducing. Some antibodies have many binding sites, not just two. These can be used to bind to a number of pathogens
- **neutralisation** — the antibody binds to an antigen, preventing it from being active

Some pathogens release toxins. One type of antibody, called an antitoxin, can bind to the toxin making it harmless.

Most pathogens need a binding site that is used to attach to a host cell before gaining entry. Some antibodies can attach to this binding site and block it. This means that the pathogen cannot bind to the host cell and enter it.

Revision activity

Add links in to your disease mind maps you started earlier. Cover how the body responds to fight the diseases.

Primary response

The first time a pathogen invades the body, it produces a **primary response**. This takes a few days as the specific B and T cells must be selected, the cells must divide and then differentiate and the antibodies must be manufactured. As a result, the peak of activity and the maximum concentration of antibodies are not achieved until several days after infection.

The **primary response** is the immune system's response to a first infection.

Secondary response

On any subsequent invasion by the same pathogen with identical antigens on its surface, the immune response is a **secondary response**. The immune system can respond more quickly and with higher intensity — it is much more effective. This is because the blood carries many memory cells that are specific to this pathogen. They divide and differentiate into plasma cells, which then manufacture antibodies. The response is quick enough to prevent the pathogen taking hold and causing symptoms of illness.

The **secondary response** is the immune system's response to a second or subsequent infection by the same pathogen.

Immunity and vaccination

Active and passive immunity

Revised

Active immunity is acquired through activation of the immune system. It involves the selection of specific lymphocytes and the production of antibodies and memory cells. The memory cells remain in the blood for a long time providing long-lasting immunity.

Passive immunity is acquired from another source. Antibodies may be injected straight into the blood or acquired from a mother's milk. These antibodies do not last long, but they do give immunity from a specific pathogen for a period of time. No memory cells are made, so the immunity is not permanent.

Active immunity is immunity that has been acquired by activation of the immune system.

Passive immunity is when someone is given antibodies produced by someone else.

Natural and artificial immunity

Revised

Natural immunity is acquired in the course of everyday activity. Natural active immunity may be the result of catching a flu virus from someone who sneezes. Natural passive immunity may be acquired through breast milk or the mother's placenta.

Artificial immunity is acquired by human intervention. Artificial active immunity may be the result of a vaccination. Artificial passive immunity results from an injection of antibodies.

Vaccination

Revised

Vaccination is the deliberate introduction of antigenic material in order to stimulate the production of antibodies. Antigenic material can take several forms:

- whole live microorganisms
- dead microorganisms
- attenuated (weakened) organisms
- a surface preparation of antigens
- a toxoid (a harmless form of a toxin)

Using live organisms is more effective than using dead ones as the live organisms can reproduce and mimic an infection better.

Examiner's tip

The discovery of vaccination using pus from milkmaids infected with cowpox is a good example of how scientific advances can have an effect on society.

Now test yourself

Tested

4 Explain why using a live organism in a vaccination is better than injecting antibodies or antigens.

Answer on p. 111

Revision activity

Add links in to your disease mind maps you started earlier.

Herd and ring vaccination

In order to be effective, vaccinations need to be used appropriately. Herd vaccination involves the systematic vaccination of all or most members of a population. This prevents the pathogen being transmitted from person to person. Ring vaccination is a response to an outbreak. All the people in the area surrounding the outbreak are vaccinated in order to prevent transmission and to isolate the outbreak in one area.

Government response to influenza

Revised

Influenza is a killer disease. History suggests that there is a major outbreak, or pandemic, of influenza every 20 or so years. This is because the influenza virus cross-breeds to produce new strains.

Health organisations such as the World Health Organization (WHO) monitor new strains of influenza in order to predict which strains might develop into a major cause for concern. The important aspects to monitor include:

- how easily the virus is transmitted
- how harmful the disease is to people who are infected
- how easily the virus may pass from other animals to humans

The response of the UK **government** to the threat of **new strains** of influenza is to prepare vaccinations to protect high-risk members of the population, including:

- people aged 65 and over
- pregnant women
- people with an ongoing serious health problem such as asthma, diabetes, heart disease or hepatitis
- people living in a long-stay residential care home
- those who are the main carer for an elderly or disabled person
- frontline health or social care workers who may be exposed to the pathogen

New medicines

Revised

Scientists are always looking for new medicines to help combat disease.

Microorganisms and **plants** produce a wide range of molecules that may be of benefit in fighting disease. Among the huge diversity of plants and microorganisms in the natural world there may be organisms that produce chemicals which are beneficial against a wide range of pathogens or in fighting diseases such as cancer. We just need to find them.

Every species of organism that is allowed to become extinct could potentially hold the cure for a major disease. It is therefore important that we try to **maintain biodiversity** and conserve as many species as possible, just in case the molecules they are capable of producing may prove to be useful.

Revision activity

Add links in to your disease mind maps you started earlier.

Cigarette smoking

The effects on the gaseous exchange system

Revised

Cigarette smoke contains many harmful chemicals. The three main components of cigarette smoke are:

- tar
- nicotine
- carbon monoxide

Tar

Tar sticks to the linings of the airways and may reach the alveoli where it increases the length of the diffusion pathway. It is made up of a combination of many chemicals which may be irritants, cause paralysis of the cilia or be **carcinogenic**.

Carcinogenic means having the potential to cause cancer.

Irritants

Irritants in the tar make the person cough more. They cause harm by stimulating the cells lining the airways:

- the goblet cells produce more mucus
- the lining of the airways becomes inflamed
- the smooth muscle in the walls of the airways becomes thickened

Together, these factors reduce the diameter of the airways, making it more difficult for air to flow along them.

Typical mistake

Many candidates describe tar lining the blood vessels. Tar does not cross the alveolus wall and does not enter the blood stream.

Chronic bronchitis

Hydrogen cyanide and carbon monoxide in cigarette smoke paralyse the cilia in the airways. This causes mucus to accumulate, producing a 'smoker's cough'. The excess mucus traps bacteria which then reproduce, causing infection. This is known as **chronic bronchitis**. The symptoms of chronic bronchitis include:

- a chronic cough
- excess mucus production
- shortness of breath
- wheezing

Emphysema

Neutrophils move into the lungs to attack the bacteria that cause infection. They use an enzyme called elastase to digest the elastic fibres in the alveoli so that they can enter the lungs. As a result, the alveoli cannot recoil during expiration. Air is trapped in the alveoli because the airways are too narrow. As the chest cavity decreases in volume during expiration, the air trapped in the alveoli causes the alveoli to burst. This reduces the surface area of the lung in a condition known as **emphysema**.

Emphysema is a disease that results in the loss of alveoli.

The symptoms of emphysema include:

- shortness of breath
- difficulty in carrying out exercise
- difficulty in expiring

Examiner's tip

This is another area in which quality of written communication can be tested — the examiner can reward candidates who write these steps in the correct sequence.

A barrel chest may also develop in people suffering from emphysema.

Emphysema and chronic bronchitis often occur together and are known as **chronic obstructive pulmonary disease (COPD)**.

Lung cancer

Certain components of tar are carcinogenic, which means that they can cause cancer. They do this by altering the genes in the cells, causing mutation. If the genes controlling cell division are altered and do not function, the cells divide in an uncontrolled way and form a tumour — this is cancer. The symptoms of **lung cancer** include:

- a persistent cough
- tumours in the lungs
- pain in the chest
- shortness of breath
- coughing up blood

The effects on the cardiovascular system

Revised

Nicotine

Nicotine is highly addictive. It crosses the wall of the alveolus and enters the blood. Once there, it has two main effects:

- it causes the platelets to become more sticky, increasing the chance of a blood clot forming
- it causes small arteries and arterioles to constrict, which reduces blood flow to the extremities and raises blood pressure

Carbon monoxide

Carbon monoxide passes across the alveolus wall to enter the blood. It combines with haemoglobin to produce carboxyhaemoglobin. This reduces the oxygen-carrying capacity of the blood. As a result, the increased heart rate can contribute to higher blood pressure, putting strain on the heart. Carbon monoxide also damages the linings of the arteries, leading to **atherosclerosis**.

Atherosclerosis, CHD and strokes

Damage to the walls of the arteries caused by carbon monoxide and high blood pressure must be repaired. The body lays down fatty substances, particularly cholesterol and fats from low density lipoproteins (LDLs). This deposition occurs just under the endothelium in the wall of the artery. It thickens the artery wall, making it less flexible and reducing the diameter of the lumen. This reduction in diameter further increases blood pressure. The reduced lumen also reduces flow of blood, and this is particularly important in the coronary arteries leading to the heart muscle. Narrowing of the coronary arteries is known as **coronary heart disease (CHD)**.

These narrowed arteries also have a rough lining that can cause blood clots to form. The narrowed lumen can trap clots that are moving in the blood, preventing blood flow in that artery. This stops the supply of oxygen to the tissues and can cause a heart attack if it is the coronary artery or a **stroke** if it is an artery leading to the brain.

Examiner's tip

This is another area in which quality of written communication can be tested — the examiner can reward candidates who write these steps in the correct sequence.

Typical mistake

Many candidates describe the deposition of fats as being 'in the artery' or 'on the wall of the artery'. Be specific — deposition is in the wall of the artery.

Revision activity

How does poor diet contribute to atherosclerosis, CHD and stroke?

Now test yourself Tested

5 Explain why people with severe atherosclerosis or other circulatory disorders may be given anticlotting drugs.

Answer on p. 111

Smoking links to disease and early death

Revised

There are two main forms of evidence that link cigarette smoking to disease and early death: **epidemiological evidence** and **experimental evidence**.

Epidemiological evidence

This is evidence gained from the study of patterns of disease:

- Lung cancer was rare in the nineteenth century before many people smoked cigarettes regularly.
- It became more common as a result of many people adopting the smoking habit.
- Almost all lung cancer patients smoke cigarettes or have done so in the past.
- Lung cancer often develops at least 20 years after smoking started.

Experimental evidence

This is evidence gained from experiments on living organisms:

- Dogs that are forced to inhale cigarette smoke often develop lung cancer.
- Mice and other animals treated with substances isolated from cigarette smoke often develop tumours.

Exam practice

1 **(a)** Describe three ways in which HIV is transmitted. [3]

(b) The following table shows the percentage of people with new HIV infections in four different parts of the world.

Year	Percentage of people with new HIV infections			
	Western Europe	**Eastern Europe**	**Far East**	**Sub-Saharan Africa**
1980	0.0	0.0	0.0	0.0
1990	0.1	0.2	0.1	2.0
2000	0.4	0.5	0.3	8.5
2010	0.3	0.9	0.3	15.2

(i) Describe the trends shown by the data in the table. [4]

(ii) Suggest why the trend in Western Europe is different from the trend in other parts of the world. [3]

(c) Discuss the advantages and disadvantages of treating HIV/AIDS patients with antibiotics. [4]

2 **(a)** Explain the difference between antigens and antibodies. [3]

(b) Describe the structure of an antibody and explain how it is adapted to its function. [4]

(c) Suggest why people in underdeveloped parts of the world such as sub-Saharan Africa may be unable to make sufficient antibodies. [2]

3 **(a)** Describe the effects of cigarette tar on the mammalian gaseous exchange system. [9]

(b) The following table shows the percentage of people who smoked regularly in the UK and the numbers of cases of lung cancer in one city.

Year	Percentage of people who smoked		Number of lung cancer patients	
	Men	**Women**	**Men**	**Women**
1900	2	0	0	0
1910	3	2	3	0
1920	72	1	3	0
1930	67	3	4	1
1940	63	34	37	2
1950	78	46	42	5
1960	73	37	39	26
1970	59	34	46	23

(i) Using the data in the table, discuss the evidence that lung cancer is linked to smoking cigarettes. [4]

(ii) When asked about the trends shown in the table, a student wrote the following comment: 'As the percentage of people who smoke rises, the number of people with lung cancer increases. It is clear from the data that smoking cigarettes causes lung cancer'. Comment on the validity of this statement. [3]

(iii) Suggest what data should be collected to improve the research and explain why it would help. [2]

Answers and quick quiz 11 online

Online

Examiner's summary

By the end of this chapter you should be able to:

- ✔ Discuss the meanings of the terms *health* and *disease*.
- ✔ Define and discuss the meanings of the terms *parasite* and *pathogen*.
- ✔ Describe the causes and transmission of malaria, HIV/AIDS and tuberculosis (TB), and discuss the global impact of these three diseases.
- ✔ Define the terms *immune response*, *antigen* and *antibody*.
- ✔ Describe the primary defences against pathogens and parasites.
- ✔ Describe the action of phagocytes.
- ✔ Describe the structure of antibodies and outline how they work.
- ✔ Describe the structure of T and B cells.
- ✔ Describe the mode of action of T and B cells and the role of memory cells.
- ✔ Compare and contrast primary and secondary immune responses.
- ✔ Compare and contrast active, passive, natural and artificial immunity.
- ✔ Explain how vaccination can control disease.
- ✔ Discuss the responses of governments to the threat of influenza.
- ✔ Outline the need for maintaining biodiversity in relation to potential new sources of medicines.
- ✔ Describe the effects of smoking on the mammalian gaseous exchange system.
- ✔ Describe the effects of nicotine and carbon monoxide on the cardiovascular system.
- ✔ Evaluate evidence linking cigarette smoking to disease and early death.

12 Biodiversity

Species, habitat and biodiversity

A **species** is a group of individuals with similar anatomy, physiology, biochemistry and behaviour, and which can interbreed to produce fertile offspring. Members of one species generally live in one habitat and show adaptations to living in that habitat.

A **habitat** is an ecological or environmental area that is inhabited by individuals from one species.

Biodiversity is the diversity of life. It includes all the different plant, animal, fungus and microorganism species in the world, the genes they contain and the ecosystems of which they form a part. It should be considered at three levels: the range of habitats within an ecosystem, the range of species within a habitat and the genetic variation within a species.

Revision activity

Draw a mind map with biodiversity in the centre. Include the three levels of biodiversity and what they mean.

Examiner's tip

This topic contains a number of key scientific terms with precise meanings. You should learn them and use them only in the correct context.

A **species** is a group of organisms with similar adaptations that live and breed together to produce fertile offspring.

A **habitat** is the home or environment of a particular species.

Biodiversity is the variety of life on Earth.

Measuring biodiversity

Species richness and evenness

Revised

A measure of the number of species in a habitat is called the **species richness**. A habitat that is dominated by just one species with only one or two individuals of each of the other species would not be considered to be biologically diverse. A habitat in which all species are equally represented is more diverse. This is known as **species evenness**. Therefore, a measurement of the biodiversity of a habitat should account for both the number of species and the number of individuals of each species living in that habitat. A diverse habitat would contain a large number of species, all of them represented by a sizeable population rather than by just one or two individuals.

Species richness is the number of different species in a habitat.

Species evenness is how evenly those species are represented throughout the habitat.

Sampling

Revised

Most habitats are fairly large in area and contain large numbers of plants and animals. It is therefore impossible to count the number of individuals in each species. **Sampling** involves studying small parts of the habitat in detail and then multiplying the results from the samples to calculate the

Now test yourself

1 Explain the significance of species evenness and species richness.

Answer on p. 111

Tested

population in the whole area. The assumption is made that the sample plots are representative of the entire habitat.

When sampling it is important to consider:

- the size of the sample areas — this depends on the size of the habitat
- the number of sample areas used — the more sample areas used the better, as the results will be more reliable
- the sampling technique used — this must be identical in every sample
- respect for habitat — the sampling should not disturb the habitat more than is essential
- that sample selection must be random (see below)

Random sampling

It is all too easy to look at a habitat and choose to sample areas that look more interesting. If the rest of the habitat is not diverse, this could lead to an estimate of biodiversity for the whole area that is too high. Also, if you concentrate on interesting-looking sample areas, you may miss many species that do not look colourful or are easily missed because they are small. This could lead to an estimate of diversity that is too low.

Typical mistake

Many candidates forget to mention that sample areas must be random or cannot explain how to make them random.

Random sampling can be achieved by:

- using a computer to generate random numbers, which are then used as coordinates to locate sample areas on an imaginary grid placed over the habitat
- carrying out systematic samples, such as every 10 metres in each direction

Now test yourself

2 Explain why sampling must be random.

Answer on p. 111

Tested ☐

Carrying out sampling

The method of sampling depends upon the type of vegetation in the habitat and what type of organisms you are trying to sample. It is important to measure both the number of species (species richness) and the number of individuals in each species (species evenness).

Sampling plants

Large plants such as trees can be counted individually. Smaller plants can be sampled using quadrats. These are square frames of a suitable size, which are placed over a random site and examined closely to identify all the plants inside the quadrat. A quantitative sample can be achieved by measuring the percentage cover of each species within the quadrat. This can be done using the following methods:

- Point sampling — place a point frame in the quadrat and count the number of examples of each species that touch each point. The total number of points touching each species can be converted to percentage frequency data (e.g. if a species touches 4 out of 10 points, it is 40% frequent).
- Grid sampling — divide the quadrat using string into a known number of smaller squares (often 100). By doing this you can generate percentage frequency data (e.g. present in 1 square = 1% frequent), as well as direct counts.

Sampling animals

Large animals can be sampled by careful observation and counting. Smaller animals will need to be caught or trapped.

Small mammals can be trapped using a humane trap such as a Longworth trap. It is possible to estimate the population size by the mark and recapture technique. This involves two separate trapping sessions. The animals caught first time are marked in a way that causes them no harm. If the number of animals trapped in first session is T1, the number caught in the second session is T2 and the number caught in the second session that are already marked is T3, the total population can be given by the formula:

$$\text{Number in population} = \text{T1} \times \frac{\text{T2}}{\text{T3}}$$

Ground-living invertebrates can be collected using a pitfall trap, whereas invertebrates in leaf litter can be collected using a Tullgren funnel. Invertebrates in trees can be collected using a stick to knock a branch and collecting the organisms that fall to the ground in a sheet. Invertebrates in grass and shrubs can be collected by sweep netting, and pond life can be sampled by netting.

Simpson's Index of Diversity

Revised

Simpson's Index of Diversity measures the probability that two individuals randomly selected from a sample will belong to the same species. There are several versions of the formula, so make sure you are consistent. The most commonly used version is:

$$D = 1 - \Sigma\,(n/N)^2$$

where n is the total number of individuals in a particular species and N is the total number of individuals in all species.

If it is not possible to count all the individual plants in an area, percentage cover can be used.

The resultant value always ranges between 0 and 1.

Examiner's tip

You may be asked to calculate Simpson's Index of Diversity from data provided. Calculations involving this index are not as difficult as many students imagine. The formula and a partly completed table are often provided.

Using the index

The data collected from a field may look like those in Table 12.1.

Table 12.1 Data collected from field sampling

Species	Percentage cover
Yorkshire fog	16
Meadow grass	74
Bent grass	2
Thistle	3
Buttercup	4
Dock	1

To measure Simpson's Index of Diversity, further calculations are carried out (Table 12.2).

Table 12.2 Calculations for Simpson's Index

Species	Percentage cover	n/N	n/N^2
Yorkshire fog	16	0.16	0.0256
Meadow grass	74	0.74	0.5476
Bent grass	2	0.02	0.0004
Thistle	3	0.03	0.0009
Buttercup	4	0.04	0.0016
Dock	1	0.01	0.0001
Total	100	1.00	0.5762

Typical mistake

Some candidates don't find calculations easy — here they often forget to square each number or to subtract from 1 as the final step in the calculation.

Using Simpson's Index of Diversity:

$$D = 1 - (0.0256 + 0.5476 + 0.0004 + 0.0009 + 0.0016 + 0.0001)$$

$$D = 1 - 0.5762$$

$$D = 0.42$$

Now test yourself Tested

3 Explain why an ecologist might want to calculate the diversity of a habitat.

Answer on p. 111

Significance and implications

A **high value** (close to 1) indicates:

- a habitat with high diversity
- there are a good number of species (high species richness)
- the species are relatively evenly represented (high species evenness)
- this sort of habitat should be stable and may survive a certain amount of disruption
- it is probably a habitat that is worth conserving

A **low value** (close to 0) indicates:

- a habitat with low diversity
- the habitat is dominated by one or a few species
- this sort of habitat may be unstable and damaged by disruption
- the habitat may be manmade

In the example above, a diversity index of 0.42 is not particularly high. This habitat is dominated by one species (meadow grass). If this species were harmed by human action, the habitat may be unstable.

Now test yourself Tested

4 Explain why a conservationist may want to conserve a more diverse habitat.

Answer on p. 111

Global biodiversity

Current estimates

Revised

Estimates of **global diversity** vary and there are many reasons for this variation:

- new species are being discovered and described
- some species are becoming extinct
- many species are yet to be discovered
- new techniques are helping scientists to define new species
- some groups of organisms are not well known (e.g. the prokaryotes may contain many more species than are currently known)
- some species may have been named twice by different people, but better communication is correcting these errors

Exam practice

1 **(a)** State what is meant by the terms *biodiversity, species richness* and *species evenness*. [3]

(b) Biodiversity can be measured at three levels: habitat, species and genetic. Explain what is meant by each level and explain the significance of high diversity in each case. [6]

(c) A student collected the following data from two fields.

Species	Percentage cover	
	Field A	Field B
Rye grass	76	14
Bent grass	0	84
Fescue	45	0
Dandelion	9	0
Buttercup	24	2
Daisy	18	0
Total	172	100

(i) Using Simpson's Index of Diversity, calculate the value of *D* for field B. [2]

(ii) The student calculated the value of *D* for field A as 0.7. Suggest what the relative values for fields A and B mean about their diversity and their value as habitats. [4]

2 **(a)** Describe how the diversity of a named habitat, such as a meadow, can be measured. [10]

(b) Suggest why an ecologist may want to measure the diversity of a habitat. [3]

Answers and quick quiz 12 online

Online

Examiner's summary

By the end of this chapter you should be able to:

- ✔ Define the terms *species, habitat* and *biodiversity*.
- ✔ Explain how biodiversity may be considered at the level of habitat, species and genetic.
- ✔ Explain the importance of sampling when measuring the biodiversity of a habitat.
- ✔ Describe how random samples can be taken.
- ✔ Describe how to measure biodiversity in a habitat.
- ✔ Use Simpson's Index of Diversity.
- ✔ Outline the significance of high and low values of Simpson's Index of Diversity.
- ✔ Discuss estimates of global biodiversity.

13 Classification

Systems for classifying organisms

Classification, phylogeny and taxonomy

Revised

Classification

People have always placed things into groups to help make sense of the world around them. Placing things into groups is known as **classification**.

Classification is the arrangement of animals and plants into groups according to their qualities or characteristics.

There are many arbitrary systems of classification that group things according to factors such as colour, which can be useful. For example, many plant identification books group plants according to the colour of their flower. This helps us to identify plants as it is easy to turn to the correct page quickly. Such a classification system is called artificial classification.

The earliest classification of living things was an artificial classification that grouped living things according to whether they lived in the air, on land or in water. However, such artificial classifications are based on human perceptions and tell us nothing about how the members of the group are biologically related. Natural classification places all the organisms that are closely related to one another in one group.

Phylogeny

Phylogeny is the evolutionary history or the evolutionary relationships between organisms and groups of organisms. Natural classification systems group living things according to their similarities. The more features shared between two organisms, the more closely they are related. Therefore, natural classification reveals the phylogeny. Phylogeny can be represented in an evolutionary tree such as Darwin's tree of life.

Phylogeny is the evolutionary relationships between organisms.

Taxonomy is the branch of science that involves classification.

Taxonomy

Taxonomy is the branch of science concerned with classification. It requires detailed study of the organism to reveal the characteristics used to classify the organism.

Examiner's tip

It can be easy to confuse the terms *classification*, *phylogeny* and *taxonomy*, so make sure you understand the differences between them.

The groups into which biologists classify living organisms are called **taxonomic groups** or taxa (singular: taxum). The taxonomic grouping system follows a **hierarchy**:

- domain
- kingdom
- phylum
- class
- order
- family
- genus
- species

So, species is the smallest group (or taxum). Similar species are placed in a genus. Similar genera are placed in a family, etc.

Now test yourself

1 Explain why the order of taxonomic groups is known as a hierarchy.

Answer on p. 111

Tested

The five kingdoms

Revised

In the **five kingdom system of classification**, the kingdoms are separated by a number of characteristic features. These are summarised in Table 13.1.

Table 13.1 The characteristic features of the five kingdoms

	Kingdom				
Feature	**Monera**	**Protoctista**	**Fungi**	**Plantae**	**Animalia**
Cellular	Unicellular	Unicellular, some multicellular (algae)	Acellular (body composed of mycelium). Yeasts are unicellular	Multicellular	Multicellular
Nucleus	No	Yes	Yes (cytoplasm is multinucleate)	Yes	Yes
Membrane-bound organelles	No	Yes	Yes	Yes	Yes
Cell wall	Yes (made of peptidoglycan)	Present in many species	Yes (made of chitin)	Yes (made of cellulose)	No
Nutrition	Autotrophic, heterotrophic or parasitic	Autotrophic, heterotrophic or parasitic	Heterotrophic	Autotrophic (photosynthetic)	Heterotrophic
Locomotion	Some have flagella	Some have an undulipodium, some have cilia	None	None	Muscular tissue

Binomial system of nomenclature

The **binomial system of nomenclature** is the way in which we use two Latin words to name each species. The first word is the name of the genus to which the species belongs and the second word is the specific or species name. For example, in the term *Homo sapiens*, *Homo* is the genus of man and *sapiens* is our species name.

The Latin names should always be written in italics or underlined. The genus name should be written with an upper case first letter, whereas the species should be in lower case (Table 13.2).

Table 13.2 Classifying three species according to taxonomic group

	Species		
Taxonomic group	**Human**	**Lion**	**Wild cat**
Kingdom	Animalia	Animalia	Animalia
Phylum	Chordata	Chordata	Chordata
Class	Mammal	Mammal	Mammal
Order	Primate	Carnivora	Carnivora
Family	Hominidae	Felidae	Felidae
Genus	*Homo*	*Panthera*	*Felis*
Species	*sapiens*	*leo*	*sylvestris*

Now test yourself

2 Explain why fungi were once classified as plants, but are no longer in this kingdom.

Answer on p. 111

Tested

Typical mistake

Candidates often use the binomial system incorrectly, forgetting to capitalise the genus name or not using italics for the genus and species names.

Dichotomous keys

Revised

A **dichotomous key**, which consists of questions with simple yes or no answers, is used to identify organisms. The questions are worded simply and ask about the visible features of the organism. Each response leads to another question or to the identity of the organism.

Examiner's tip

A dichotomous key should have simple questions asking about the presence or absence of features or the relative size of features. The number of questions should be one less than the number of organisms named in the key.

Typical mistake

Don't write too many questions in a dichotomous key or include questions that are ambiguous, such as 'Does it have large leaves?'. Whether the observer thinks the leaves are large or not is based on a comparison with other leaves, so it would be better to write such questions 'Are the leaves longer than 10 cm?'.

Revision activities

Choose a few small plants and write a dichotomous key to help someone identify them.

Modern classification

Carl Linnaeus set up the first true system of classification, which is still used today. He used **observable features** of the organisms and grouped them according to the number of similarities present.

Since then, taxonomists have studied organisms in ever more detail, describing all of their observable features. They then decide if any differences observed are simply a variation within the species or significant enough to place the organism in a different species.

Recent approaches

Revised

More **recent approaches** involving sequencing of DNA and proteins have also been used. This **molecular evidence** provides a genetic and biochemical comparison, which are thought to be more accurate than comparing observable features. Those organisms with similar DNA sequences must be closely related. Since changes in DNA sequence are caused by mutation and mutations are random, it is assumed that more mutations means a longer time has elapsed since the species became separate. Therefore, the more mutations or differences in the DNA, the less closely related the species.

As the DNA sequence is used to produce proteins, it is equally valid for studying amino acid sequences in proteins. The best proteins are those that occur in all living things, such as proteins associated with respiration and protein synthesis. These include cytochrome C and RNA polymerase.

Once the taxonomic groups are constructed on the basis of genetic similarities, they will automatically reveal the evolutionary **relationships between organisms**.

Examiner's tip

Classification systems have changed over recent years and this is a great opportunity to test certain aspects of How Science Works — in particular the idea that scientific knowledge changes continually as research reveals new evidence.

Now test yourself

Tested

3 Explain why using DNA sequencing to classify organisms automatically matches their phylogeny.

Answer on p. 111

The three domains

Revised

Until recent years, the five kingdom system of classification was universally accepted. However, recent research has revealed that this hierarchy may not be accurate. This research included detailed study of the sequence of bases in the RNA of the ribosomes.

One kingdom, the Monera, consists of a wide diversity of prokaryotic organisms. However, part of this kingdom is significantly different from the rest of the kingdom. This led to the idea that the kingdom should be divided into two major groups or domains: the Bacteria and the Archaea. The Bacteria are fundamentally different from all other living things. The Archaea are still prokaryotic, but they are more similar to the Eukaryotes.

The **three domain system of classification** of Bacteria, Archaea and Eukaryota is now widely accepted and the domains are a taxonomic level above the kingdoms.

Exam practice

1 The leopard, *Panthera pardus*, is a member of the cat family. Complete the following table to show its full classification. [5]

Kingdom	
	Chordata
Class	Mammalia
	Carnivora
Family	Felidae
Genus	
	pardus

2 **(a)** Wild flower identification books group plants according to the colour of their flowers. Explain why this is artificial classification. [2]

(b) State three features that can be used to classify a mammal into the kingdom Animalia. [3]

(c) **(i)** List three features of organisms belonging to the kingdom Fungi. [3]

(ii) State two features that fungi share with plants. [2]

3 Explain the significance of research into the sequence of bases in DNA from different species. [7]

Answers and quick quiz 13 online

Online

Examiner's summary

By the end of this chapter you should be able to:

- ✔ Define the terms *classification*, *phylogeny* and *taxonomy*.
- ✔ Explain the relationship between classification and phylogeny.
- ✔ Describe the classification of species into the taxonomic hierarchy.
- ✔ Outline the characteristic features of the five kingdoms.
- ✔ Outline the use of the binomial system of nomenclature.
- ✔ Use a dichotomous key.
- ✔ Discuss the classification of organisms using observable features and molecular evidence.
- ✔ Compare and contrast the five kingdom and three domain classification systems.

14 Evolution

Variation

The term **variation** refers to the differences between individuals.

Variation can occur between members of the **same species**. These differences could be:

- simple observable features such as colour
- biochemical differences such as the precise sequence of amino acids in a protein
- behavioural differences such as the type of food eaten

These differences are usually relatively minor such as differences in size, weight or hair colour, or they could be more obvious such as the differences between the sexes.

Variation can also occur between members of **different species**. This depends on how closely related one species is to the other:

- If the species are closely related, such as the lion and the tiger, the differences may not be great.
- If the species are not closely related, the differences will be greater.

Examiner's tip

Remember that variation is the key to evolution — variation must occur before any characteristic can become beneficial and selected for.

Continuous and discontinuous variation

Revised

Continuous variation

Continuous variation is seen where there are no distinct groups or categories. There is a full range between two extremes. This form of variation is caused by:

- a number of genes interacting together
- the environment

Examples of continuous variation include height and body weight. The continuously variable feature can be quantified and data are usually presented in the form of a histogram.

Discontinuous variation

Discontinuous variation is seen where there are distinct groups or categories and there are no in-between types. This type of variation is usually caused by one gene (or possibly a small number of genes). The discontinuously variable feature cannot be quantified — it is qualitative.

Examples of discontinuous variation include gender and possession of resistance or immunity. Data are usually presented in the form of a bar chart.

Continuous variation is variation that shows a complete range with no distinct groups.

Discontinuous variation is variation that produces distinct groups.

Typical mistake

Students tend to plot bar charts with no spaces between the bars, but there should be gaps between them.

Revision activities

Write a list of features that are continuously variable and a separate list of features that are discontinuously variable.

Causes of variation

Revised

There are two **causes of variation**:

- genetic
- environmental

Many variable features may be affected by both causes. For example, skin colour in humans is genetically determined. However, exposure to the sun will cause extra pigment production, causing the skin to tan.

Genetic

Genetic causes of variation are a result of differences in the sequence of bases in the DNA. They are caused by mutations that arise spontaneously and randomly, and are passed on from one generation to the next. They usually cause discontinuous variation. Examples include:

- number of limbs
- eye colour
- ability to roll the tongue

Environmental

Environmental causes of variation are caused by variations in exposure to certain environmental conditions. They are not passed from one generation to the next and cause continuous variation. Examples include:

- skin colour caused by exposure to sunlight
- body mass

Now test yourself

1 Explain how genetic differences cause visible variation between members of the same species.

Answer on p. 111

Tested

Revision activities

Write a list of features that show variation caused by genes, a separate list of features that show variation caused by the environment and a third list of features that show variation caused by both genes and the environment.

Adaptations of organisms to their environments

All members of a species possess similar **adaptations**, which enable the species to survive and thrive in their **environments**. These adaptations can be categorised according to whether they are behavioural, physiological or anatomical.

Adaptations are features that help organisms to survive in their habitats.

Behavioural adaptations

Revised

Behavioural adaptations are often associated with feeding, nesting and mating:

- Male deer will compete for dominance in the rutting season. This allows one dominant male to father all the offspring.
- Robins usually choose a nest site in a hole in a tree stump or wall a few inches above the ground. This means they are not competing with other bird species.
- In dry conditions some plants open their stomata to make the leaves wilt. This reduces the surface area exposed to hot sun and decreases the rate of transpiration.
- Yeasts will form spores when conditions are not favourable. This helps individual cells to survive during poor conditions.

Physiological adaptations

Revised

Physiological adaptations are those associated with how the body systems function:

- Fish pass water over their gills in one direction, as opposed to mammals which have a tidal flow of air into their lungs. This is because water is much more dense and difficult to slow down and stop so that the direction of movement can be reversed.
- The kidneys of mammals extract water from the urine before excreting nitrogenous waste. This helps to reduce the need to find and drink water.
- Some plants, called C4 plants, use an unusual way of collecting (fixing) carbon dioxide at night. This means they can keep their stomata closed during the hottest part of the day, reducing loss of water via transpiration, but are still able to photosynthesise.
- Yeasts will respire anaerobically when there is no oxygen in their habitat. This means they can produce ATP and continue to grow.

Anatomical adaptations

Revised

Anatomical adaptations are those to do with structure:

- Predators have sharp teeth to help kill and chew their prey. They also have a strong jaw joint so that it does not become dislocated by a struggling victim.
- Herbivores have a long and complex digestive system. This allows them to digest plant tissues, which are much more difficult to digest than animal tissues.
- Plants have long, deep roots with many root hairs. This enables them to absorb water and minerals from the soil.
- Some plants such as the black mangrove, which lives in waterlogged soil, grow roots up into the air. This allows the roots to gain oxygen from the air above the anaerobic soil.

Typical mistake

Students often forget to explain how each adaptation helps survival of the species.

Examiner's tip

In examination questions, you are likely to be given some information about a particular species and its habitat. The question will then ask you to explain how the features described help the organism to survive.

Revision activities

Which of the adaptations listed are unusual? Select one from each category and explain why it is unusual.

Natural selection and evolution

Darwin's theory of natural selection

Revised

Darwin made **four observations** in proposing his theory:

- Parents tend to produce more offspring than are able to survive.
- Populations tend to remain a constant size.
- Offspring look similar to their parents.
- No two individuals look identical.

From the above observations, Darwin made the following **deductions**:

- There must be a struggle to exist.
- Some offspring are better adapted to surviving than others.

From the above deductions, Darwin reached the following **consequences**:

- Parents produce too many offspring.
- There must be competition to survive.
- Not all the offspring survive.
- Although all offspring look similar to their parents, some are better adapted to their environment.
- The better adapted offspring survive.
- Those that survive will go on to reproduce.
- The less well adapted offspring do not survive.
- The better adapted individuals pass on their features to the next generation.

Darwin called this theory 'evolution by **natural selection**'.

The theory of **natural selection** is Charles Darwin's theory that certain inherited traits which increase an organism's chances of survival are favoured over less beneficial traits.

Variation, adaptation and selection

Darwin's theory of evolution by natural selection is widely accepted. Species are thought to adapt to their environment and, once well adapted, they no longer change unless the environment changes.

Members of a species show **variation**. If the environment changes, the organisms living there are no longer fully adapted, which places a **selective** pressure on the species. Some individuals show **adaptations** that enable them to survive better than others, giving them a selective advantage. These individuals survive to reproduce offspring and, over several generations, there is an increase in the proportion of individuals showing the advantageous characteristics — the species is evolving.

Typical mistake

Students sometimes forget to mention that for new strains or new species to evolve, variation, adaptation and selection must occur over several generations.

Now test yourself

2 Draw a flow diagram to explain the process of evolution by natural selection.

Answer on p. 111

Tested

Speciation

If the species evolves enough to be recognised as different from its ancestors then a new species has been formed — this is **speciation**. Speciation can also occur when two populations evolve separately — once they can no longer interbreed they are separate species.

Speciation is the formation of a new species.

Evidence for the theory of evolution

Revised

There are many lines of evidence for the theory of **evolution**. These include fossils and molecular evidence in the form of the structure of DNA and other molecules, particularly proteins.

Fossils

Darwin used a lot of evidence from **fossils** to back up his theory of natural selection. Fossils take two forms:

- imprints of ancient organisms
- the remains of organisms that have died and become mineralised

Fossils are found in sedimentary rocks. They are formed when an organism leaves an imprint in soft mud, or actually dies and comes to rest in the mud. As the mud hardens to form rock, the imprint or body remains in the rock. Fossils formed relatively recently are near the surface of modern rocks, whereas those formed many millions of years ago are found deeper below the surface.

Scientists study fossils in minute detail and carefully describe their anatomy and morphology. Similarities between fossils can be used to reveal evolutionary relationships (phylogeny).

Molecular evidence

Certain chemicals, such as **DNA**, proteins and RNA, are universal to all living things. Variation, in part, is caused by changes (mutations) in the DNA, which produces changes in proteins. As members of the same species have similar DNA, they will also have similar proteins. As evolution occurs and one species becomes two, the DNA will accumulate more changes, as will the structure of the proteins it codes for. Therefore, closely related species will have similar DNA and proteins, but more distantly related species will have DNA and proteins with more differences.

Sequencing the bases in DNA and the amino acids in proteins shows the similarities and differences between species. This reveals their evolutionary relationships in the same way that similarities in anatomy and morphology reveal evolutionary relationships.

Molecular evidence is, perhaps, more reliable and more convincing than that of fossils.

Evolution in recent times

Revised

Pesticide resistance

When **pesticides** are used, they kill all susceptible insects. However, some individuals may have a degree of **resistance**. These few individuals may survive and breed to pass on their resistance to successive generations. The pesticide has acted as a selective agent. As successive generations show some variation, it is possible for the insects to become increasingly resistant to higher and higher concentrations of pesticide. The implications are that insects can damage food crops and with no effective pesticides we will be unable to prevent this damage.

Drug resistance

Most bacteria are susceptible to antibiotics. However, some **microorganisms** may show some resistance to **drugs**. Just as with insect resistance to pesticides, the bacterial species eventually evolves resistance. The implications are that we have developed only a limited number of antibiotics and, once bacteria have evolved resistance to them all, we will have no further defences to help us combat disease.

Typical mistake

Some candidates make the mistake of referring to the 'evolution of immunity' to pesticides or antibiotics.

Examiner's tip

Questions sometimes take the form of a description of the mechanism of evolution. This is likely to be in the context of artificial selection of farm animals or plants.

Now test yourself Tested

3 Explain how strains of the bacterium *Chlostridium difficile* can become resistant to antibiotics.

Answer on p. 112

Exam practice

1 **(a) (i)** State two features of continuous variation. [2]

(ii) State two features of discontinuous variation. [2]

(b) List the causes of variation. [2]

(c) Explain what is meant by the term *selection*. [3]

2 **(a)** Explain how inappropriate use of antibiotics has given rise to the so-called superbug methicillin-resistant *Staphylococcus aureus* (MRSA). [5]

(b) Explain how the structure of proteins can be used as evidence for evolution. [4]

Answers and quick quiz 14 online

Online

Examiner's summary

By the end of this chapter you should be able to:

- ✔ Define the term *variation*.
- ✔ Discuss variation within as well as between species.
- ✔ Describe the differences between continuous and discontinuous variation.
- ✔ Explain both genetic and environmental causes of variation.
- ✔ Outline the adaptations of organisms to their environments.
- ✔ Explain the consequences drawn by Darwin in proposing his theory of natural selection.
- ✔ Define the term *speciation*.
- ✔ Outline the roles of variation, adaptation and selection in evolution.
- ✔ Discuss the fossil, DNA and molecular evidence for evolution.
- ✔ Discuss the evolution of pesticide resistance in insects and drug resistance in microorganisms.

15 Maintaining biodiversity

Reasons for conservation

Many species have become rare or endangered, often because of human activity. The **conservation** of animal and plant species is important for a variety of reasons, many of which can be applied generally to most endangered species.

Examiner's tip

The reasons for conservation are generic, but questions in the examination are likely to ask about a specific case. Be ready to apply these generic ideas to the case under investigation.

Economic reasons

Revised

Economic reasons for conserving species include the following:

- All endangered species are part of an ecosystem.
- Growth of food and timber relies on the correct functioning of ecosystems.
- Pollination of many crops relies on insects, particularly bees.
- Natural predators to pests reduce the need for pesticides.
- As yet unknown species may contain molecules that are effective medicines.
- Natural selection has already solved many problems that doctors and engineers could learn from.

Ecological reasons

Revised

Many species are in decline because of habitat destruction — finely balanced ecosystems may be disrupted by a small change. The balance of life on Earth is maintained by the activity of species within ecosystems and we do not know what knock-on effects the loss of one species may have. Ecosystems that function correctly carry out many essential **ecological** processes:

- fixing of energy from sunlight
- regulation of the oxygen and carbon dioxide levels in the atmosphere
- fresh water purification and retention
- soil formation
- maintenance of soil fertility
- mineral recycling
- waste detoxification and recycling

Ethical reasons

Revised

Many people believe that it is not right that other species become extinct as a result of human actions and for our benefit. Such extinctions may affect local people as much as the wild animals and plants. **Ethical** arguments include the following:

- Some species may have religious significance to certain people.
- Some people still live in the environment — as part of the ecosystem — and should be allowed to continue living that way.

- All life depends on functioning ecosystems.
- There is a limited supply of resources and these should be shared by all living things.
- Humans are a part of nature and should abide by its rules.
- Human intervention in nature could bring about unexpected consequences.

Typical mistake

Many students write long, heartfelt responses about the right of all organisms to live and how humankind should not 'play God'. This sort of response may gain some credit, but will not gain more than 1 or 2 marks.

Aesthetic reasons

Revised

There are several **aesthetic** reasons why animal and plant species should be conserved:

- Everyone enjoys nature or its benefits in one way or another.
- A healthy, well-balanced ecosystem, with its variety of life forms, colours and activity, is complex and beautiful.
- Being surrounded by natural systems relieves stress and helps recovery from injury.

Typical mistake

Many candidates can answer questions on why conservation is important in generic terms, but tend to get thrown by specific examples.

The importance of maintaining biodiversity

The consequences of global climate change

Revised

There are several **consequences** of **global climate change** on the **biodiversity** of plants and animals:

- Global climate change will affect the conditions in which organisms live.
- Many species will no longer be adapted to the environments where they live.
- Water levels, water temperatures and nutrient recycling will all be affected. As a result, many species may struggle to survive. These species must adapt or migrate, otherwise they face extinction.
- The migration of plants and animals will affect human populations:
  - staple food crops may no longer grow
  - **agricultural practices** will need to change
  - new pests, **diseases** and parasites will appear
- If species cannot migrate, biodiversity will decline. This loss of biodiversity means there will be less genetic variety, so natural populations will be less likely to evolve and it will become difficult to selectively breed new farm animals.

Examiner's tip

The effects of climate change could well be tested as part of a question on adaptation to the environment or evolution.

Revision activity

Draw a mind map with global warming in the centre. Include the effects of global warming on biodiversity.

The benefits for agriculture

Revised

There are several **benefits** for **agriculture** of maintaining the **biodiversity** of animal and plant species, including:

- plenty of variety for scientific research
- new strains for selective breeding
- genes that may provide crop resistance to drought, specific diseases, fungal and insect attack, and freezing

Examiner's tip

The benefits of maintaining biodiversity could be tested as part of a question about food production.

Now test yourself Tested

1 Explain why it is important that we conserve a variety of strains of crop plants.

Answer on p. 112

Approaches to conservation

Conservation *in situ* and *ex situ*

Revised

Conservation of endangered species can take place either *in situ* or *ex situ*:

- *In situ* conservation involves conserving a species in its natural habitat by creating nature reserves or preserves.
- *Ex situ* conservation involves conserving a species using controlled habitats away from its normal environment. **Botanic gardens**, **seed banks** and wildlife parks all keep groups of individuals of endangered species.

Examiner's tip

The principles of conservation are straightforward. Therefore, questions about conservation are likely to be applied to a specific case. Learn to apply your knowledge to new contexts.

Tables 15.1 and 15.2 outline the advantages and disadvantages of both forms of conservation.

Table 15.1 Advantages and disadvantages of conservation *in situ*

Advantages	Disadvantages
The organisms are in their normal environment	It can be difficult to monitor the organisms and ensure they are healthy
The habitat is conserved along with all the other species living in it	The environmental factors that caused the decline in numbers may still be present
The organisms will behave normally	Poaching or hunting may continue
It generates work for local people looking after the reserve	There may be food shortages
Ecological tourism can generate income	Disease will be difficult to treat
	Predators can be difficult to control

Table 15.2 Advantages and disadvantages of conservation *ex situ*

Advantages	Disadvantages
Research is easy	The organisms are living in an unnatural habitat
The organisms can be monitored to ensure they remain healthy	They may not behave as normal
They can be kept in separate populations to ensure a disease does not affect the whole species	They may not breed
Breeding can be controlled to prevent in-breeding and subsequent loss of genetic diversity	There is little point in conserving individuals if their natural habitat is lost and there is nowhere for them to return to
Genetic diversity can be increased by sharing specimens with other conservation sites	
Once populations have been increased, individuals can be reintroduced to the wild	
Seed banks can store the seeds of millions of **rare plant** species or plant species that are **extinct** in the wild, using the same area necessary for only a few hundred adult plants	

Now test yourself Tested

2 Explain why it is important to keep more than one population of an endangered species.

Answer on p. 112

Revision activity

Visit a local wildlife park to find out what activities are carried out for international conservation.

International cooperation

Revised

CITES

CITES (the **Convention on International Trade in Endangered Species**) is an international agreement between governments, to which countries adhere voluntarily. Its aim is to ensure that international trade in specimens of wild animals and plants does not threaten their survival.

The Rio Convention on Biological Diversity

The **Rio Convention on Biological Diversity** recognises that people need to secure resources of food, water and medicines. However, it promotes development that is sustainable and partner countries agree to adopt *ex situ* conservation measures with shared resources.

Now test yourself

3 Explain why international cooperation is essential for successful conservation.

Answer on p. 112

Tested

Environmental impact assessments

Revised

An **environmental impact assessment (EIA)** is an assessment of the possible positive or negative impacts that a proposed project may have on the environment, including environmental, social and economic aspects. They are a legal requirement before all major building developments can proceed. The developer's EIA must include **estimates** of **biodiversity** and the effect of the development on the biodiversity of the area. Local planning authorities can set certain constraints on the development in order to reduce the expected impact.

Revision activity

Consider a proposed development in your area. What sort of environmental impact will it have? (If you have no proposed development, consider a recent development.)

Exam practice

1 The mountain gorilla (*Gorilla beringei beringei*) is an endangered species. There are thought to be about 790 individuals left. These live in two populations, one in the Virunga mountains of central Africa and the other in southwest Uganda.

(a) Explain why such a low population puts the species at risk. [3]

(b) **(i)** Suggest how local people could help conserve these gorillas. [3]

(ii) Suggest how *ex situ* conservation techniques could be used to help the gorillas. [3]

2 Explain why seed banks are considered to be more important than botanical gardens. [2]

3 **(a)** What does the acronym CITES mean? [1]

(b) **(i)** Describe the aims of CITES. [2]

(ii) Suggest two ways in which these aims can be achieved. [2]

4 Describe the role of an environmental impact assessment in local planning. [3]

Answers and quick quiz 15 online

Online

Examiner's summary

By the end of this chapter you should be able to:

- ✔ Outline the reasons for the conservation of animal and plant species.
- ✔ Discuss the consequences of global climate change on the biodiversity of plants and animals.
- ✔ Explain the benefits for agriculture of maintaining the biodiversity of animal and plant species.
- ✔ Describe the conservation of endangered plant and animal species, both *in situ* and *ex situ*.
- ✔ Discuss the role of botanic gardens in the *ex situ* conservation of rare plant species or plant species extinct in the wild, with reference to seed banks.
- ✔ Discuss the importance of international cooperation in species conservation with reference to CITES and the Rio Convention on Biological Diversity.
- ✔ Discuss the significance of environmental impact assessments for local authority planning decisions.

Now test yourself answers

Chapter 1

1. The resolution of a light microscope is not sufficient to separate objects that are closer than 200 nm. Therefore, an image magnified at greater than 1500× contains no extra detail.
2. A scanning electron microscope image is in three-dimensions and is a surface view. A transmission electron microscope can see details inside organelles.
3. $I = AM$, $A = \frac{I}{M}$
4. There are 1000 nanometers (nm) in one micrometer (µm).
5. The organelles could be cut in a different plane. Organelles, especially mitochondria, can change shape and be larger in more active cells.

Chapter 2

1. Active transport (using energy from ATP) of molecules or ions across a membrane.
2. Compartments can specialise — they can keep all the molecules and enzymes associated with one process together where they are in higher concentration and are not interfered with by other processes.
3. The phospholipid bilayer stops movement of ions and polar molecules. Transport proteins (carrier proteins and channel proteins) have attachment sites for specific molecules.
4. A signal molecule may be specific to a certain target cell. It also has a specific shape. If the receptor has a specific shape that is complementary to the signal molecule, the signal molecule will be able to bind only to that receptor site and will only affect that particular type of target cell.
5. The phospholipid bilayer is impermeable to charged molecules and ions. They cannot diffuse through the bilayer, so they must have an alternative route. This is through proteins that span the membrane, so it is known as facilitated diffusion.
6. Proteins are large molecules and cannot easily diffuse through the phospholipid bilayer. Also, they will not fit through protein pores.
7. The cell wall is strong and can withstand the pressure of water in the cell.
8. A strong salt solution has a low water potential. The cell will have a higher water potential. As water moves down the water potential gradient from inside the cell to outside, the cell loses water. Water will no longer push against the sides of the cell and it will lose turgidity.
9. The solution from outside the cell will fill the gap between the plasma membrane and the wall. The wall is fully permeable so the solution simply moves in through the wall.

Chapter 3

1. Mitosis produces two daughter cells that are genetically identical to the original cell. It must make a copy of the DNA so that there is one full set of genes to enter each daughter cell, otherwise the cells of a multicellular organism would not contain the same DNA and genes.
2. Cell division and growth are energy-requiring processes. Each daughter cell must have enough energy to grow. Each daughter cell must also have sufficient organelles to carry out the requirements of a living cell.
3. Cells divide and then differentiate. Differentiation turns off some of the genes so that they are not expressed. The cell than changes shape and modifies its contents to be able to carry out its function — this is called specialisation.
4. Many cells that are all specialised in the same way form a tissue. They cannot do much on their own as they are so small. However, a large number of cells all performing the same task can achieve much more together. For example, one muscle cell could not move an organism, but many muscle cells together can generate enough force to move a limb.

Chapter 4

1. Size, surface area to volume ratio and level of activity.
2. An amoeba is small, it has a large surface area to volume ratio and can absorb all the oxygen it needs through its surface. A large tree has a small surface area to volume ratio, so its surface area is too small to supply the volume. Also, diffusion is too slow to allow oxygen to reach all parts of the plant.
3. Concentration on the supply side (needs to be high), concentration on the demand side (needs to be low), thickness of barrier (needs to be thin).
4. To prevent air entering or leaving the system through the nose. This would alter the volume of air in the chamber and lead to inaccurate readings.
5. As the subject exhales, carbon dioxide enters the chamber. A high concentration of carbon dioxide alters the breathing rate and may become harmful. Soda lime absorbs the carbon dioxide. This causes the volume in the air chamber to decrease and allows the volume of oxygen used to be calculated.

Chapter 5

1. Double circulation ensures that the oxygenated blood is separated from the deoxygenated blood. This makes transport of oxygen more efficient. Double circulation also allows a higher pressure in the systemic side so that materials can be delivered further and more quickly. The lower pressure in the pulmonary side reduces the chance of damage to the capillaries in the lungs.

2 The volume of the ventricle chambers is the same on both sides of the heart. The wall of the left ventricle has much thicker muscle so that it can create a higher pressure to push the blood further around the body.
3 The valves are pushed open and closed by changes in pressure in the blood. When the atrium wall contracts and the pressure in the atrium is higher than in the ventricle, the valves are pushed open as the blood flows from higher to lower pressure. When the ventricle walls contract, the pressure in the ventricles rises above that in the atria and the valves are pushed closed as the blood tries to flow away from the high pressure area.
4 The blood flowing from the atrium to the ventricle moves more slowly than the electrical stimulus. The time delay allows for the blood to flow and fill the ventricle before the ventricle walls are stimulated to contract.
5 The pressure in the ventricle rises above the pressure in the atrium. Blood starts to move back towards the atrium, but it is trapped in the atrioventricular valve and pushes the valve closed.
6 Artery walls are thick and contain thick layers of muscle and elastic tissue. They also contain collagen. The collagen is strong and prevents the artery wall bursting under high pressure. The inner lining of the artery wall is folded and can unfold as the pressure and volume of blood increase so that the lining does not get damaged or split. As the pressure and volume of blood increase, the wall of the artery stretches to accommodate the extra blood. Once the pressure starts to decrease, the elastic fibres recoil to reduce the diameter of the artery again.
7 Veins have much less smooth muscle and elastic tissue in their walls. They do not need to stretch and recoil as the pressure is never high. They also have much less collagen. The veins are often flattened and only become round in cross-section when they are full of blood. They also contain valves in their walls which prevent the blood flowing backwards away from the heart.
8 Air contains only about 20% oxygen. As the oxygen in the air is used up, the oxygen tension decreases and the blood in the lungs is less able to extract oxygen from the air. With less saturation of the haemoglobin, less oxygen is transported around the body, so less oxygen is supplied to the brain and respiring tissues.

Chapter 6

1 When water evaporates from the surfaces of the cells in the leaf, it creates a pull or tension on the chain of water molecules in the xylem. This suction effect reduces the pressure in the xylem vessels and they may collapse under the pressure from surrounding tissues. The lignin strengthens the walls to support them against this collapse. It is important that water can continue in an unbroken chain all the way up the xylem. If a xylem vessel becomes blocked or damaged, water can pass through the pits from one vessel to another to maintain the unbroken chain. Also, the pits allow water to pass out of the vessel into surrounding cells and tissues.
2 Companion cells carry out the active processes that are used to load sucrose into the sieve tube elements.
3 Increasing the temperature increases evaporation, so the water potential in the air spaces of the leaf increases. This increases the water vapour potential gradient between the leaf air spaces and outside the leaf, so water vapour diffuses out of the leaf more quickly. Higher light intensity causes the stomata to open wider to allow more carbon dioxide to enter the leaf for photosynthesis. With wider stomatal openings, more water vapour can diffuse out of the leaf.
4 Transpiration is the loss of water vapour from the aerial parts of the plant. The transpiration stream replaces this loss by transporting water into the roots and up the stem to the leaves.
5 In spring when the leaf buds are opening, they require a supply of sucrose. This is used in respiration to supply the energy needed for growth and also as a building block to make new cells. The leaf is therefore a sink. Once the leaf is formed and can start to photosynthesise, it manufactures sugars which will be transported away to other parts of the plant. The leaf is then a source.

Chapter 7

1 A $\delta+$ charge is a small charge not equivalent to a whole electron or proton charge. It is created by polarity in the molecule. Certain atoms attract electrons more strongly than others and if electrons are attracted away from one part of the molecule it leaves a $\delta+$ charge.
2 Specific heat capacity is a measure of how much energy is needed to warm up a substance. The energy is used to make the molecules move about more. Latent heat capacity is a measure of how much energy is needed to convert a substance from one state (e.g. liquid) to another (e.g. gas). The energy is used in breaking the bonds that hold the molecules together.
3 Water has a high specific heat capacity, which means that a lot of energy is needed to warm up a body of water. As living things are mostly water, this means that a body remains at a fairly constant temperature. Latent heat is the heat needed to evaporate water or sweat. Evaporating water off plant leaves and sweat off the skin helps to keep living things cool.
4 Proteins are chains of amino acids joined by peptide bonds. A peptide bond forms between the amino end of one amino acid and the carboxylic acid end of another. Each amino acid has just one amino group and one carboxylic acid group. Therefore, it cannot form an extra bond to create a branch.
5 A polypeptide is a long chain of amino acids joined by peptide bonds. A protein may consist of one polypeptide chain or it may have several polypeptide chains. A protein may also contain a non-protein prosthetic group such as the haem group in haemoglobin.
6 Metabolically active proteins need a specific shape which is achieved by the three-dimensional tertiary structure. This gives hormones a binding site and enzymes an active site.
7 The three-dimensional shape of globular proteins is usually very specific. If it is altered by changes in temperature, the protein becomes inactive.
8 Amylose is a long chain of glucose. Glucose holds energy. Amylose is large and therefore insoluble, so it does not affect

the water potential of the cell. The molecule is highly coiled, so it does not take up much space. Amylose can be broken down easily by hydrolysis to release the glucose units when energy is needed.

9 Cellulose is a large molecule. It has many hydroxyl groups that can form hydrogen bonds and these bind to other cellulose molecules, making fibrils.

10 Lipids are high energy compounds. The fatty acid chains can be broken down and enter respiration to release energy. Lipids are not water soluble, so they do not affect the water potential of the cell.

11 Phospholipids have a phosphate group attached to the 'head' end. This is hydrophilic and will mix with water — it will twist towards water. The fatty acid 'tails' are hydrophobic and will not mix with water — they twist away from water. A group of phospholipid molecules will orientate so that they form a globule with the heads on the outside in contact with the water and the tails on the inside acting as a barrier to the water.

12 A solution with low concentration of glucose contains less glucose than one with high concentration. If the Benedict's reagent is in excess, the low concentration of glucose will react with only some of the available Benedict's reagent, leaving some remaining. If there is a lot of glucose present, it will react with more of the Benedict's reagent, leaving less or none remaining.

Chapter 8

1 DNA is genetic material — it carries the genetic code from one generation to the next. It must be stable so that the code in the genes remains the same. This means that when it is copied to pass on to other cells, it remains the same. Therefore, all the cells in one individual will have the same codes and those codes can be passed on accurately to any offspring.

2 If any errors were made in the copying of the base sequence, the gene code would be changed. This is a mutation. A mutation may not produce the required protein or the protein may not function properly.

Chapter 9

1 You can imagine the need for activation energy as a boulder sitting in a hollow at the top of a hill. The boulder will not roll down the hill until it is pushed up and out of the hollow. Once over the lip of the hollow, the boulder can then roll easily down the hill. An enzyme can provide an alternative route for the reaction — removing the lip of the hollow.

2 The shape of the active site depends upon interactions between the *R* groups in the enzyme molecule. These interact through bonds such as ionic bonds between oppositely charged parts of the molecule and disulfide bonds between sulphur atoms. There are also weaker hydrogen bonds and interactions such as hydrophilic and hydrophobic interactions and forces of attraction and repulsion between charges. As a substrate molecule enters the active site, the physical presence of the molecule can interfere with the weaker interactions in the active site. Also, any charges on the substrate molecule will interact with charges in the active site and this can cause the shape of the active site to change slightly.

3 A non-competitive inhibitor binds to a part of the enzyme molecule away from the active site. The presence of the inhibitor modifies the pattern of bonds and interactions within the enzyme molecule. As a result, the active site changes shape and is no longer complementary to the shape of the substrate. The substrate cannot enter the active site and no enzyme substrate complexes are formed.

4 In any practical procedure involving enzymes, there may be many possible variables. Each variable has the potential to alter the activity of the enzyme and therefore change the rate of reaction. If two or more variables are changed at the same time, both may have an effect on the rate of reaction. It is not possible to distinguish between the effects that each variable has on the rate of reaction unless the experimenter can be certain that only one variable has been changed.

5 Variables such as temperature and pH need to be controlled so that they do not change during the practical test and affect the rate of reaction. A control is a separate test set up to demonstrate that a particular factor is required for the reaction to occur. It is usual to set up a control in which the enzyme, or source of enzyme, is missing to show that the enzyme is essential for the reaction to occur.

Chapter 10

1 There are many factors that contribute to the development of CHD. No single factor will cause the disease on its own. Each factor increases the chances that CHD will develop.

2 High blood pressure stretches the arteries. Every time the heart beats, it creates a pulse of high pressure that travels along the arteries. The arteries stretch to accommodate this pulse. If the pressure is too high, it stretches the artery too much. This could split the inner lining of the artery (the endothelium). This will expose the receptors for LDLs.

3 Select a number of cows that have a high milk yield. Select a bull that comes from a mother with low fat (less creamy) milk. Breed them together to produce offspring. Select the offspring with the best combination of high yield and low fat milk. Continue selecting and breeding for many generations until the desired combination is achieved.

4 When the animals get ill, their growth slows down. If there are antibiotics in the food, they will prevent the growth of any bacteria or fungi that infect the animals, so the animals should not get ill. This allows the animals to grow more quickly. The antibiotics are not effective unless the animals are infected. The antibiotics may harm the bacteria living in the animals' digestive system and reduce their ability to digest their food. Bacteria and fungi can develop resistance to antibiotics. Overuse will help this resistance to develop.

5 Bacteria and fungal spores are everywhere. If food is not packaged, it will become re-infected quickly. Even if the food has been treated by cooking or irradiation, it may not be able to prevent the bacteria and fungi growing. Some treatments have a lasting effect. For example, salting,

smoking and pickling should keep the food from spoiling even if it is not packaged.

Chapter 11

1 Malaria is a disease that causes ill health to the patient. The disease is caused by a pathogen, a microorganism called *Plasmodium*. *Plasmodium* lives in the blood of the victim and feeds on the haemoglobin in the red blood cells. This means that the victim's blood is less able to transport oxygen. The pathogen is transmitted by a parasite that survives by feeding on human blood. The parasite is the mosquito *Anopheles*, and it is the female *Anopheles* that feeds on human blood. A pathogen can be transmitted by the female *Anopheles* biting an infected person and becoming infected itself, before biting another person who is not infected.

2 The NHS no longer vaccinates everyone against TB. Many migrants from less well developed parts of the world live in inner cities where they hope to find work. Often these migrants live in crowded conditions that allow the bacterium to be transmitted. Some of these migrants may be infected with the *Mycobacterium* bacterium, which provide a source for transmission. HIV reduces resistance to TB.

3 All cells have proteins or glycoproteins on their cell surface membranes. On the surface of a pathogen, these may have functions such as enabling the cell to bind to other cells or to receptor sites on the host cell membrane. The immune system can recognise these proteins or glycoproteins and use them as antigens to recognise foreign cells.

4 Injecting antibodies will provide instant immunity but it is short term — the antibodies do not last long in the body. No memory cells are made, so there is no long-term immunity. If dead cells or antigens are injected, this allows the immune system to become activated and produce antibodies itself along with memory cells. These memory cells are what provide long-term immunity. However, there will be only a limited number of pathogen cells or antigens and the immune system may not be fully activated. If living cells are injected, they will reproduce and increase in number — they mimic a real infection. This activates the immune system more fully and provides complete immunity.

5 Atherosclerosis leaves a roughened surface on the lining of the arteries. This may affect the flow of blood and cause an increased chance of clotting. Any circulatory disorder that reduces blood flow may increase the chance of clotting. If clots form in the blood, they may get caught in areas where the blood vessels are narrowed. This causes reduced blood flow and a shortage of oxygen to the affected tissue.

Chapter 12

1 Species evenness is a measure of how many individuals there are of each species. If all the species in a habitat are represented by a large number of individuals or a large population, the population is likely to be stable. It also means that the habitat is more diverse. Species richness is a measure of how many different species are found in a habitat. If there are a lot of species, the habitat is more diverse and likely to be more stable.

2 Any bias in selecting sites for samples will make the results less valid. This is because the element of choice can be affected by what the sampler sees. If he or she chooses an area with fewer species, the overall measurement of diversity will be an underestimate. If he or she chooses an area with more species, the overall estimate will be an overestimate. It would also be tempting to select areas that contain rare species, which would affect the decision about how important the habitat is as an environment for rare organisms.

3 Monitoring diversity over a number of years can reveal trends and changes in the populations of organisms. Careful monitoring with comparisons to physical factors on the site can help to gain an understanding of the environmental factors that affect the growth and survival of organisms. If the site is likely to be affected by a development, it may be important to establish if there are rare or endangered species in the area in order to ascertain if the site is of importance for conservation.

4 Diverse habitats contain more species and are more stable and likely to survive some level of disruption. When resources are limited and only certain areas can be conserved, it is more important to conserve a habitat containing more species rather than one that contains few species or is unstable and could be damaged despite the attempts to conserve it.

Chapter 13

1 All members of one species belong to the same genus, which in turn belongs to the same family, order, class, etc. As you move up the hierarchy, the taxonomic groups get bigger and bigger.

2 Fungi have a spreading body structure like plants and are not able to move around like plants. They also have nuclei and membrane-bound organelles like plants. However, fungi do not possess chloroplasts and are not able to photosynthesise — they are not autotrophic.

3 If the DNA of two species is similar, there have been few mutations to make changes in the DNA. As mutations are spontaneous and random, there will be more mutations and more differences in the DNA between two species that have been evolving separately for a long time, i.e. they are not closely related. If there are few differences in the DNA, the species have not been separate for a long time and are therefore closely related.

Chapter 14

1 Genes code for the structure of proteins and for the sequence of amino acids in the protein. The protein contributes to the formation of a particular feature. If the base sequence in the DNA changes, the code also changes. This alters the sequence of amino acids in the protein. The new protein may not work or may contribute to the production of a different visible feature.

2 There is a change in the environment → which causes selective pressure → due to natural variation between members of the species → some individuals possess a

feature that gives them an advantage → these individuals survive more easily → they reproduce successfully to pass on their genes.

3 A patient infected with *Chlostridium difficile* is given antibiotics to kill the bacterium. Due to natural variation, some of the bacteria have some resistance. If the whole course of antibiotics is not completed, some of the resistant bacteria will survive. These reproduce and pass on the gene for resistance to future generations. The next generation also shows variation and some individuals are more resistant than others. Over many generations of bacterial growth and reproduction, successive generations will become increasingly more resistant to the antibiotic.

Chapter 15

1 Each variation has a genetic basis. The genes provide characteristics that are adaptations to the local environment. If conditions change and farmers can no longer grow the same strains, new strains must be found. Selective breeding using high-yield strains crossed with those that can grow well in the prevailing conditions may produce viable crops. Equally, if a new disease or fungus starts to cause reduced yields, breeding with a resistant strain could produce a high-yield resistant strain that grows well locally.

2 Keeping two or more populations of a species allows independent evolution that may enhance genetic variation. Occasional cross-breeding between the populations reduces the chances of harmful genetic combinations arising through inbreeding. If one population is harmed by a disease, the second population ensures that the whole species is not wiped out. Having two populations also benefits scientific research into the species as comparisons can be made between the two populations.

3 Wild animals and plants do not respect international boundaries. If one country spends a lot of effort on conserving a species, the species does not benefit if another country allows the wholesale slaughter of that species. If trade in an endangered species is allowed in one country, this encourages poaching in the natural habitat of another country that is trying to conserve the species.